DATA S☑ W9-BLP-210

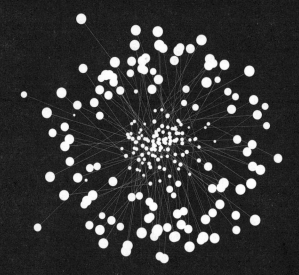

The MIT Press Essential Knowledge Series

DATA SCIENCE

JOHN D. KELLEHER
AND BRENDAN TIERNEY

The MIT Press | Cambridge, Massachusetts | London, England

This book was set in Chaparral Pro by Toppan Best-set Premedia Limited. Printed and bound in the United States of America.

Library of Congress Cataloging-in-Publication Data

Names: Kelleher, John D., 1974- author. | Tierney, Brendan, 1970- author.
Title: Data science / John D. Kelleher and Brendan Tierney.
Description: Cambridge, MA : The MIT Press, [2018] | Series: The MIT Press essential knowledge series | Includes bibliographical references and index.
Identifiers: LCCN 2017043665 | ISBN 9780262535434 (pbk. : alk. paper)
Subjects: LCSH: Big data. | Machine learning. | Data mining. | Quantitative research.
Classification: LCC QA76.9.B45 K45 2018 | DDC 005.7--dc23 LC record available at https://lccn.loc.gov/2017043665

10   9   8   7   6   5

CONTENTS

SERIES FOREWORD

The MIT Press Essential Knowledge series offers accessible, concise, beautifully produced pocket-size books on topics of current interest. Written by leading thinkers, the books in this series deliver expert overviews of subjects that range from the cultural and the historical to the scientific and the technical.

In today's era of instant information gratification, we have ready access to opinions, rationalizations, and superficial descriptions. Much harder to come by is the foundational knowledge that informs a principled understanding of the world. Essential Knowledge books fill that need. Synthesizing specialized subject matter for nonspecialists and engaging critical topics through fundamentals, each of these compact volumes offers readers a point of access to complex ideas.

Bruce Tidor
Professor of Biological Engineering and Computer Science
Massachusetts Institute of Technology

The goal of data science is to improve decision making by basing decisions on insights extracted from large data sets. As a field of activity, data science encompasses a set of principles, problem definitions, algorithms, and processes for extracting nonobvious and useful patterns from large data sets. It is closely related to the fields of data mining and machine learning, but it is broader in scope. Today, data science drives decision making in nearly all parts of modern societies. Some of the ways that data science may affect your daily life include determining which advertisements are presented to you online; which movies, books, and friend connections are recommended to you; which emails are filtered into your spam folder; what offers you receive when you renew your cell phone service; the cost of your health insurance premium; the sequencing and timing of traffic lights in your area; how the drugs you may need were designed; and which locations in your city the police are targeting.

The growth in use of data science across our societies is driven by the emergence of big data and social media, the speedup in computing power, the massive reduction in the cost of computer memory, and the development of more powerful methods for data analysis and modeling, such as deep learning. Together these factors mean that it has never been easier for organizations to gather, store, and process data. At the same time, these technical innovations and the broader application of data science means that the ethical challenges related to the use of data and individual privacy have never been more pressing. The aim of this book is to provide an introduction to data science that covers the essential elements of the field at a depth that provides a principled understanding of the field.

Chapter 1 introduces the field of data science and provides a brief history of how it has developed and evolved. It also examines why data science is important today and some of the factors that are driving its adoption. The chapter finishes by reviewing and debunking some of the myths associated with data science. Chapter 2 introduces fundamental concepts relating to data. It also describes the standard stages in a data science project: business understanding, data understanding, data preparation, modeling, evaluation, and deployment. Chapter 3 focuses on data infrastructure and the challenges posed by big data and the integration of data from multiple sources. One aspect of a typical data infrastructure that can be challenging is that data in databases and data warehouses often reside on servers different from the servers used for data analysis. As a consequence, when large data sets are handled, a surprisingly large amount of time can be spent moving data between the servers a database or data warehouse are living on and the servers used for data analysis and machine learning. Chapter 3 begins by describing a typical data science infrastructure for an organization and some of the emerging solutions to the challenge of moving large data sets within a data infrastructure, which include the use of in-database machine learning, the use of Hadoop for data storage and processing, and the development of hybrid database systems that seamlessly combine traditional database software and Hadoop-like solutions. The chapter concludes by highlighting some of the challenges in integrating data from across an organization into a unified representation that is suitable for machine learning. Chapter 4 introduces the field of machine learning and explains some of the most popular machine-learning algorithms and models, including neural networks, deep learning, and decision-tree models. Chapter 5 focuses on linking machine-learning expertise with real-world problems by reviewing a range of standard business problems and describing how they can be solved by machine-learning solutions. Chapter 6 reviews the ethical implications of data science, recent developments in data regulation, and some of the new computational approaches to preserving the privacy of individuals within the data science process. Finally, chapter 7 describes some of the areas where data science will have a significant impact in the near future and sets out some of the principles that are important in determining whether a data science project will succeed.

ACKNOWLEDGMENTS

John and Brendan thank Paul McElroy and Brian Leahy for reading and commenting on early drafts. They also thank the two anonymous reviewers who provided detailed and helpful feedback on the manuscript and the staff at the MIT Press for their support and guidance.

John thanks his family and friends for their support and encouragement during the preparation of this book and dedicates this book to his father, John Bernard Kelleher, in recognition of his love and friendship.

Brendan thanks Grace, Daniel, and Eleanor for their constant support while he was writing yet another book (his fourth), juggling the day jobs, and traveling.

WHAT IS DATA SCIENCE?

Data science encompasses a set of principles, problem definitions, algorithms, and processes for extracting non-obvious and useful patterns from large data sets. Many of the elements of data science have been developed in related fields such as machine learning and data mining. In fact, the terms *data science*, *machine learning*, and *data mining* are often used interchangeably. The commonality across these disciplines is a focus on improving decision making through the analysis of data. However, although data science borrows from these other fields, it is broader in scope. Machine learning (ML) focuses on the design and evaluation of algorithms for extracting patterns from data. Data mining generally deals with the analysis of structured data and often implies an emphasis on commercial applications. Data science takes all of these considerations into account but also takes up other challenges, such as the capturing, cleaning, and transforming of unstructured social media and web data; the use of big-data technologies to store and process big, unstructured data sets; and questions related to data ethics and regulation.

Using data science, we can extract different types of patterns. For example, we might want to extract patterns that help us to identify groups of customers exhibiting similar behavior and tastes. In business jargon, this task is known as *customer segmentation*, and in data science terminology it is called *clustering*. Alternatively, we might want to extract a pattern that identifies products that are frequently bought together, a process called *association-rule mining*. Or we might want to extract patterns that identify strange or abnormal events, such as fraudulent insurance claims, a process known as *anomaly* or *outlier detection*. Finally, we might want to identify patterns that help us to classify things. For example, the following rule illustrates what a classification pattern extracted from an email data set might look like: *If an email contains the phrase "Make money easily," it is likely to be a spam email.* Identifying these types of classification rules is known as *prediction*. The word *prediction* might seem an odd choice because the rule doesn't predict what will happen in the future: the email already is or isn't a spam email. So it is best to think of prediction patterns as predicting the missing value of an attribute rather than as predicting

If a human expert can easily create a pattern in his or her own mind, it is generally not worth the time and effort of using data science to "discover" it.

the future. In this example, we are predicting whether the email classification attribute should have the value "spam" or not.

Although we can use data science to extract different types of patterns, we always want the patterns to be both nonobvious and useful. The example email classification rule given in the previous paragraph is so simple and obvious that if it were the only rule extracted by a data science process, we would be disappointed. For example, this email classification rule checks only one attribute of an email: Does the email contain the phrase "make money easily"? If a human expert can easily create a pattern in his or her own mind, it is generally not worth the time and effort of using data science to "discover" it. In general, data science becomes useful when we have a large number of data examples and when the patterns are too complex for humans to discover and extract manually. As a lower bound, we can take a large number of data examples to be defined as more than a human expert can check easily. With regard to the complexity of the patterns, again, we can define it relative to human abilities. We humans are reasonably good at defining rules that check one, two, or even three attributes (also commonly referred to as *features* or *variables*), but when we go higher than three attributes, we can start to struggle to handle the interactions between them. By contrast, data science is often applied in contexts where we want to look for patterns among tens, hundreds, thousands, and, in extreme cases, millions of attributes.

The patterns that we extract using data science are useful only if they give us insight into the problem that enables us to do something to help solve the problem. The phrase *actionable insight* is sometimes used in this context to describe what we want the extracted patterns to give us. The term *insight* highlights that the pattern should give us relevant information about the problem that isn't obvious. The term *actionable* highlights that the insight we get should also be something that we have the capacity to use in some way. For example, imagine we are working for a cell phone company that is trying to solve a customer *churn* problem—that is, too many customers are switching to other companies. One way data science might be used to address this problem is to extract patterns from the data about previous customers that allow us to identify current customers who are churn risks and then contact these customers and try to persuade them to stay with us. A pattern that enables us to identify likely churn customers is useful to us only if (*a*) the patterns identify the customers early enough that we have enough time to contact them before they churn and (*b*) our company is able to put a team in place to contact them. Both of these things are required in order for the company to be able to act on the insight the patterns give us.

A Brief History of Data Science

The term *data science* has a specific history dating back to the 1990s. However, the fields that it draws upon have a much longer history. One thread in this longer history is the history of data collection; another is the history of data analysis. In this section, we review the main developments in these threads and describe how and why they converged into the field of data science. Of necessity, this review introduces new terminology as we describe and name the important technical innovations as they arose. For each new term, we provide a brief explanation of its meaning; we return to many of these terms later in the book and provide a more detailed explanation of them. We begin with a history of data collection, then give a history of data analysis, and, finally, cover the development of data science.

A History of Data Gathering

The earliest methods for recording data may have been notches on sticks to mark the passing of the days or poles stuck in the ground to mark sunrise on the solstices. With the development of writing, however, our ability to record our experiences and the events in our world vastly increased the amount of data we collected. The earliest form of writing developed in Mesopotamia around 3200 BC and was used for commercial record keeping. This type of record keeping captures what is known as *transactional data*. Transactional data include event information such as the sale of an item, the issuing of an invoice, the delivery of goods, credit card payment, insurance claims, and so on. *Nontransactional data*, such as demographic data, also have a long history. The earliest-known censuses took place in pharaonic Egypt around 3000 BC. The reason that early states put so much effort and resources into large data-collection operations was that these states needed to raise taxes and armies, thus proving Benjamin Franklin's claim that there are only two things certain in life: death and taxes.

In the past 150 years, the development of the electronic sensor, the digitization of data, and the invention of the computer have contributed to a massive increase in the amount of data that are collected and stored. A milestone in data collection and storage occurred in 1970 when Edgar F. Codd published a paper explaining the *relational data model*, which was revolutionary in terms of setting out how data were (at the time) stored, indexed, and retrieved from databases. The relational data model enabled users to extract data from a database using simple queries that defined what data the user wanted without requiring the user to worry about the underlying structure of the data or where they were physically stored. Codd's paper provided the foundation for modern databases and the development of *structured query language* (SQL), an international standard for defining database queries. Relational databases store data in tables with a structure of one row per instance and one column per attribute. This structure is ideal for storing data because it can be decomposed into natural attributes.

Databases are the natural technology to use for storing and retrieving structured transactional or *operational* data (i.e., the type of data generated by a company's day-to-day operations). However, as companies have become larger and more automated, the amount and variety of data generated by different parts of these companies have dramatically increased. In the 1990s, companies realized that although they were accumulating tremendous amounts of data, they were repeatedly running into difficulties in analyzing those data. Part of the problem was that the data were often stored in numerous separate databases within the one organization. Another difficulty was that databases were optimized for storage and retrieval of data, activities characterized by high volumes of simple operations, such as SELECT, INSERT, UPDATE, and DELETE. In order to analyze their data, these companies needed technology that was able to bring together and reconcile the data from disparate databases and that facilitated more complex analytical data operations. This business challenge led to the development of *data warehouses*. In a data warehouse, data are taken from across the organization and integrated, thereby providing a more comprehensive data set for analysis.

Over the past couple of decades, our devices have become mobile and networked, and many of us now spend many hours online every day using social technologies, computer games, media platforms, and web search engines. These changes in technology and how we live have had a dramatic impact on the amount of data collected. It is estimated that the amount of data collected over the five millennia since the invention of writing up to 2003 is about 5 exabytes. Since 2013, humans generate and store this same amount of data *every day*. However, it is not only the amount of data collected that has grown dramatically but also the variety of data. Just consider the following list of online data sources: emails, blogs, photos, tweets, likes, shares, web searches, video uploads, online purchases, podcasts. And if we consider the metadata (data describing the structure and properties of the raw data) of these events, we can begin to understand the meaning of the term *big data*. Big data are often defined in terms of the three Vs: the extreme *volume* of data, the *variety* of the data types, and the *velocity* at which the data must be processed.

The advent of big data has driven the development of a range of new database technologies. This new generation of databases is often referred to as "*NoSQL databases*." They typically have a simpler data model than traditional relational databases. A NoSQL database stores data as objects with attributes, using an object notation language such as the *JavaScript Object Notation* (JSON). The advantage of using an object representation of data (in contrast to a relational table-based model) is that the set of attributes for each object is encapsulated within the object, which results in a flexible representation. For example, it may be that one of the objects in the database, compared to other objects, has only a subset of attributes. By contrast, in the standard tabular data structure used by a relational database, all the data points should have the same set of attributes (i.e., columns). This flexibility in object representation is important in contexts where the data cannot (due to variety or type) naturally be decomposed into a set of structured attributes. For example, it can be difficult to define the set of attributes that should be used to represent free text (such as tweets) or images. However, although this representational flexibility allows us to capture and store data in a variety of formats, these data still have to be extracted into a structured format before any analysis can be performed on them.

The existence of big data has also led to the development of new data-processing frameworks. When you are dealing with large volumes of data at high speeds, it can be useful from a computational and speed perspective to distribute the data across multiple servers, process queries by calculating partial results of a query on each server, and then merge these results to generate the response to the query. This is the approach taken by the *MapReduce* framework on Hadoop. In the MapReduce framework, the data and queries are mapped onto (or distributed across) multiple servers, and the partial results calculated on each server are then reduced (merged) together.

A History of Data Analysis

Statistics is the branch of science that deals with the collection and analysis of data. The term *statistics* originally referred to the collection and analysis of data about the state, such as demographics data or economic data. However, over time the type of data that statistical analysis was applied to broadened so that today statistics is used to analyze all types of data. The simplest form of statistical analysis of data is the summarization of a data set in terms of *summary (descriptive) statistics* (including measures of a central tendency, such as the *arithmetic mean*, or measures of variation, such as the *range*). However, in the seventeenth and eighteenth centuries the work of people such as Gerolamo Cardano, Blaise Pascal, Jakob Bernoulli, Abraham de Moivre, Thomas Bayes, and Richard Price laid the foundations of probability theory, and through the nineteenth century many statisticians began to use probability distributions as part of their analytic tool kit. These new developments in mathematics enabled statisticians to move beyond descriptive statistics and to start doing *statistical learning*. Pierre Simon de Laplace and Carl Friedrich Gauss are two of the most important and famous nineteenth-century mathematicians, and both made important contributions to statistical learning and modern data science. Laplace took the intuitions of Thomas Bayes and Richard Price and developed them into the first version of what we now call *Bayes' Rule*. Gauss, in his search for the missing dwarf planet Ceres, developed the *method of least squares*, which enables us to find the best model that fits a data set such that the error in the fit minimizes the sum of squared differences between the data points in the data set and the model. The method of least squares provided the foundation for statistical learning methods such as *linear regression* and *logistic regression* as well as the development of *artificial neural network* models in artificial intelligence (we will return to least squares, regression analysis, and neural networks in chapter 4).

Between 1780 and 1820, around the same time that Laplace and Gauss were making their contributions to statistical learning, a Scottish engineer named William Playfair was inventing statistical graphics and laying the foundations for modern *data visualization* and *exploratory data analysis*. Playfair invented the *line chart* and *area chart* for time-series data, the *bar chart* to illustrate comparisons between quantities of different categories, and the *pie chart* to illustrate proportions within a set. The advantage of visualizing quantitative data is that it allows us to use our powerful visual abilities to summarize, compare, and interpret data. Admittedly, it is difficult to visualize large (many data points) or complex (many attributes) data sets, but data visualization is still an important part of data science. In particular, it is useful in helping data scientists explore and understand the data they are working with. Visualizations can also be useful to communicate the results of a data science project. Since Playfair's time, the variety of data-visualization graphics has steadily grown, and today there is research ongoing into the development of novel approaches to visualize large, multidimensional data sets. A recent development is the *t-distributed stochastic neighbor embedding* (t-SNE) algorithm, which is a useful technique for reducing high-dimensional data down to two or three dimensions, thereby facilitating the visualization of those data.

The developments in probability theory and statistics continued into the twentieth century. Karl Pearson developed modern hypothesis testing, and R. A. Fisher developed statistical methods for *multivariate analysis* and introduced the idea of *maximum likelihood estimate* into statistical inference as a method to draw conclusions based on the relative probability of events. The work of Alan Turing in the Second World War led to the invention of the electronic computer, which had a dramatic impact on statistics because it enabled much more complex statistical calculations. Throughout the 1940s and subsequent decades, a number of important computational models were developed that are still widely used in data science. In 1943, Warren McCulloch and Walter Pitts proposed the first mathematical model of a *neural network*. In 1948, Claude Shannon published "A Mathematical Theory of Communication" and by doing so founded *information theory*. In 1951, Evelyn Fix and Joseph Hodges proposed a model for *discriminatory analysis* (what would now be called a *classification* or *pattern-recognition* problem) that became the basis for modern *nearest-neighbor models*. These postwar developments culminated in 1956 in the establishment of the field of *artificial intelligence* at a workshop in Dartmouth College. Even at this early stage in the development of artificial intelligence, the term *machine learning* was beginning to be used to describe programs that gave a computer the ability to learn from data. In the mid-1960s, three important contributions to ML were made. In 1965, Nils Nilsson's book titled *Learning Machines* showed how neural networks could be used to learn linear models for classification. The following year, 1966, Earl B. Hunt, Janet Marin, and Philip J. Stone developed the concept-learning system framework, which was the progenitor of an important family of ML algorithms that induced decision-tree models from data in a top-down fashion. Around the same time, a number of independent researchers developed and published early versions of the *k-means* clustering algorithm, now the standard algorithm used for data (customer) segmentation.

The field of ML is at the core of modern data science because it provides algorithms that are able to automatically analyze large data sets to extract potentially interesting and useful patterns. Machine learning has continued to develop and innovate right up to the present day. Some of the most important developments include *ensemble models*, where predictions are made using a set (or committee) of models, with each model voting on each query, and *deep-learning neural networks*, which have multiple (i.e., more than three) layers of neurons. These deeper layers in the network are able to discover and learn complex attribute representations (composed of multiple, interacting input attributes that have been processed by earlier layers), which in turn enable the network to learn patterns that generalize across the input data. Because of their ability to learn complex attributes, deep-learning networks are particularly suitable to high-dimensional data and so have revolutionized a number of fields, including *machine vision* and *natural-language processing*.

As we discussed in our review of database history, the early 1970s marked the beginning of modern database technology with Edgar F. Codd's relational data model and the subsequent explosion of data generation and storage that led to the development of data warehousing in the 1990s and more recently to the phenomenon of big data.

However, well before the emergence of big data, in fact by the late 1980s and early 1990s, the need for a field of research specifically targeting the analysis of these large data sets was apparent. It was around this time that the term *data mining* started to be used in the database communities. As we have already discussed, one response to this need was the development of data warehouses. However, other database researchers responded by reaching out to other research fields, and in 1989 Gregory Piatetsky-Shapiro organized the first workshop on *knowledge discovery in databases* (KDD). The announcement of the first KDD workshop neatly sums how the workshop focused on a multidisciplinary approach to the problem of analyzing large databases:

> Knowledge discovery in databases poses many interesting problems, especially when databases are large. Such databases are usually accompanied by substantial domain knowledge which can significantly facilitate discovery. Access to large databases is expensive—hence the need for sampling and other statistical methods. Finally, knowledge discovery in databases can benefit from many available tools and techniques from several different fields including expert systems, machine learning, intelligent databases, knowledge acquisition, and statistics.[1]

In fact, the terms *knowledge discovery in databases* and *data mining* describe the same concept, the distinction being that data mining is more prevalent in the business communities and KDD more prevalent in academic communities. Today, these terms are often used interchangeably,[2] and many of the top academic venues use both. Indeed, the premier academic conference in the field is the International Conference on Knowledge Discovery and Data Mining.

The Emergence and Evolution of Data Science

The term *data science* came to prominence in the late 1990s in discussions relating to the need for statisticians to join with computer scientists to bring mathematical rigor to the computational analysis of large data sets. In 1997, C. F. Jeff Wu's public lecture "Statistics = Data Science?" highlighted a number of promising trends for statistics, including the availability of large/complex data sets in massive databases and the growing use of computational algorithms and models. He concluded the lecture by calling for statistics to be renamed "data science."

In 2001, William S. Cleveland published an action plan for creating a university department in the field of data science (Cleveland 2001). The plan emphasizes the need for data science to be a partnership between mathematics and computer science. It also emphasizes the need for data science to be understood as a multidisciplinary endeavor and for data scientists to learn how to work and engage with subject-matter experts. In the same year, Leo Breiman published "Statistical Modeling: The Two Cultures" (2001). In this paper, Breiman characterizes the traditional approach to statistics as a data-modeling culture that views the primary goal of data analysis as identifying the (hidden) stochastic data model (e.g., *linear regression*) that explains how the data were generated. He contrasts this culture with the algorithmic-modeling culture that focuses on using computer algorithms to create prediction models that are accurate (rather than explanatory in terms of how the data was generated). Breiman's distinction between a statistical focus on models that explain the data versus an algorithmic focus on models that can accurately predict the data highlights a core difference between statisticians and ML researchers. The debate between these approaches is still ongoing within statistics (see, for example, Shmueli 2010). In general, today most data science projects are more aligned with the ML approach of building accurate prediction models and less concerned with the statistical focus on explaining the data. So although data science became prominent in discussions relating to statistics and still borrows methods and models from statistics, it has over time developed its own distinct approach to data analysis.

Since 2001, the concept of data science has broadened well beyond that of a redefinition of statistics. For example, over the past 10 years there has been a tremendous growth in the amount of the data generated by online activity (online retail, social media, and online entertainment). Gathering and preparing these data for use in data science projects has resulted in the need for data scientists to develop the programming and hacking skills to scrape, merge, and clean data (sometimes unstructured data) from external web sources. Also, the emergence of big data has meant that data scientists need to be able to work with big-data technologies, such as Hadoop. In fact, today the role of a data scientist has become so broad that there is an ongoing debate regarding how to define the expertise and skills required to carry out this role.[3] It is, however, possible to list the expertise and skills that most people would agree are relevant to the role, which are shown in figure 1. It is difficult for an individual to master all of these areas, and, indeed, most data scientists usually have in-depth knowledge and real expertise in just a subset of them. However, it is important to understand and be aware of each area's contribution to a data science project.

Data scientists should have some domain expertise. Most data science projects begin with a real-world, domain-specific problem and the need to design a data-driven solution to this problem. As a result, it is important for a data scientist to have enough domain expertise that they understand the problem, why it is important, and how a data science solution to the problem might fit into

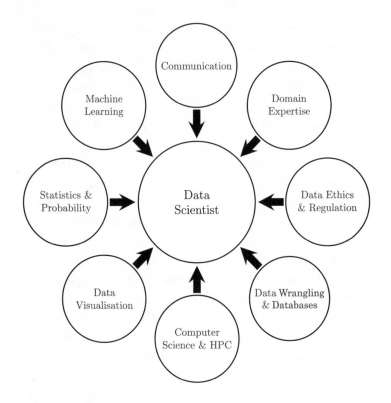

Figure 1 A skills-set desideratum for a data scientist.

an organization's processes. This domain expertise guides the data scientist as she works toward identifying an optimized solution. It also enables her to engage with real domain experts in a meaningful way so that she can illicit and understand relevant knowledge about the underlying problem. Also, having some experience of the project domain allows the data scientist to bring her experiences from working on similar projects in the same and related domains to bear on defining the project focus and scope.

Data are at the center of all data science projects. However, the fact that an organization has access to data does not mean that it can legally or should ethically use the data. In most jurisdictions, there is antidiscrimination and personal-data-protection legislation that regulates and controls the use of data usage. As a result, a data scientist needs to understand these regulations and also, more broadly, to have an ethical understanding of the implications of his work if he is to use data legally and appropriately. We return to this topic in chapter 6, where we discuss the legal regulations on data usage and the ethical questions related to data science.

In most organizations, a significant portion of the data will come from the databases in the organization. Furthermore, as the data architecture of an organization grows, data science projects will start incorporating data from a variety of other data sources, which are commonly referred to as "big-data sources." The data in these data sources can exist in a variety of different formats, generally a database of some form—relational, NoSQL, or Hadoop. All of the data in these various databases and data sources will need to be integrated, cleansed, transformed, normalized, and so on. These tasks go by many names, such as *extraction, transformation, and load*, "data munging," "data wrangling," "data fusion," "data crunching," and so on. Like source data, the data generated from data science activities also need to be stored and managed. Again, a database is the typical storage location for the data generated by these activities because they can then be easily distributed and shared with different parts of the organization. As a consequence, data scientists need to have the skills to interface with and manipulate data in databases.

A range of computer science skills and tools allows data scientists to work with big data and to process it into new, meaningful information. *High-performance computing* (HPC) involves aggregating computing power to deliver higher performance than one can get from a stand-alone computer. Many data science projects work with a very large data set and ML algorithms that are computationally expensive. In these situations, having the skills required to access and use HPC resources is important. Beyond HPC, we have already mentioned the need for data scientists to be able to scrap, clean, and integrate web data as well as handle and process unstructured text and images. Furthermore, a data scientist may also end up writing in-house applications to perform a specific task or altering an existing application to tune it to the data and domain being processed. Finally, computer science skills are also required to be able to understand and develop the ML models and integrate them into the production or analytic or back-end applications in an organization.

Presenting data in a graphical format makes it much easier to see and understand what is happening with the data. Data visualization applies to all phases of the data science process. When data are inspected in tabular form, it is easy to miss things such as outliers or trends in distributions or subtle changes in the data through time. However, when data are presented in the correct graphical form, these aspects of the data can pop out. Data visualization is an important and growing field, and we recommend two books, *The Visual Display of Quantitative Information* by Edward Tufte (2001) and *Show Me the Numbers: Designing Tables and Graphs to Enlighten* by Stephen Few (2012) as excellent introductions to the principles and techniques of effective data visualization.

Methods from statistics and probability are used throughout the data science process, from the initial gathering and investigation of the data right through to the comparing of the results of different models and analyses produced during the project. Machine learning involves using a variety of advanced statistical and computing techniques to process data to find patterns. The data scientist who is involved in the applied aspects of ML does not have to write his own versions of ML algorithms. By understanding the ML algorithms, what they can be used for, what the results they generate mean, and what type of data particular algorithms can be run on, the data scientist can consider the ML algorithms as a gray box. This allows him to concentrate on the applied aspects of data science and to test the various ML algorithms to see which ones work best for the scenario and data he is concerned with.

Finally, a key aspect of being a successful data scientist is being able to communicate the story in the data. This story might uncover the insight that the analysis of the data has revealed or how the models created during a project fit into an organization's processes and the likely impact they will have on the organization's functioning. There is no point executing a brilliant data science project unless the outputs from it are used and the results are communicated in such a way that colleagues with a nontechnical background can understand them and have confidence in them.

Where Is Data Science Used?

Data science drives decision making in nearly all parts of modern societies. In this section, we describe three case studies that illustrate the impact of data science: consumer companies using data science for sales and marketing; governments using data science to improve health, criminal justice, and urban planning; and professional sporting franchises using data science in player recruitment.

Data Science in Sales and Marketing

Walmart has access to large data sets about its customers' preferences by using point-of-sale systems, by tracking customer behavior on the Walmart website, and by tracking social media commentary about Walmart and its products. For more than a decade, Walmart has been using data science to optimize the stock levels in stores, a well-known example being when it restocked strawberry Pop-Tarts in stores in the path of Hurricane Francis in 2004 based on an analysis of sales data preceding Hurricane Charley, which had struck a few weeks earlier. More recently, Walmart has used data science to drive its retail revenues in terms of introducing new products based on analyzing social media trends, analyzing credit card activity to make product recommendations to customers, and optimizing and personalizing customers' online experience on the Walmart website. Walmart attributes an increase of 10 to 15 percent in online sales to data science optimizations (DeZyre 2015).

The equivalent of up-selling and cross-selling in the online world is the "recommender system." If you have watched a movie on Netflix or purchased an item on Amazon, you will know that these websites use the data they collect to provide suggestions for what you should watch or buy next. These recommender systems can be designed to guide you in different ways: some guide you toward blockbusters and best sellers, whereas others guide you toward niche items that are specific to your tastes. Chris Anderson's book *The Long Tail* (2008) argues that as production and distribution get less expensive, markets shift from selling large amounts of a small number of hit items to selling smaller amounts of a larger number of niche items. This trade-off between driving sales of hit or niche products is a fundamental design decision for a recommender system and affects the data science algorithms used to implement these systems.

Governments Using Data Science

In recent years, governments have recognized the advantages of adopting data science. In 2015, for example, the US government appointed Dr. D. J. Patil as the first chief data scientist. Some of the largest data science initiatives spearheaded by the US government have been in health. Data science is at the core of the Cancer Moonshot[4] and Precision Medicine Initiatives. The Precision Medicine Initiative combines human genome sequencing and data science to design drugs for individual patients. One part of the initiative is the All of Us program,[5] which is gathering environment, lifestyle, and biological data from more than one million volunteers to create the world's biggest data sets for precision medicine. Data science is also revolutionizing how we organize our cities: it is used to track, analyze, and control environmental, energy, and transport systems and to inform long-term urban planning (Kitchin 2014a). We return to health and smart cities in chapter 7 when we discuss how data science will become even more important in our lives over the coming decades.

The US government's Police Data Initiative[6] focuses on using data science to help police departments understand the needs of their communities. Data science is also being used to predict crime hot spots and recidivism. However, civil liberty groups have criticized some of the uses of data science in criminal justice. In chapter 6, we discuss the privacy and ethics questions raised by data science, and one of the interesting factors in this discussion is that the opinions people have in relation to personal privacy and data science vary from one domain to the next. Many people who are happy for their personal data to be used for publicly funded medical research have very different opinions when it comes to the use of personal data for policing and criminal justice. In chapter 6, we also discuss the use of personal data and data science in determining life, health, car, home, and travel insurance premiums.

Data Science in Professional Sports

The movie *Moneyball* (Bennett Miller, 2011), starring Brad Pitt, showcases the growing use of data science in modern sports. The movie is based on the book of the same title (Lewis 2004), which tells the true story of how the Oakland A's baseball team used data science to improve its player recruitment. The team's management identified that a player's on-base percentage and slugging percentage statistics were more informative indicators of offensive success than the statistics traditionally emphasized in baseball, such as a player's batting average. This insight enabled the Oakland A's to recruit a roster of undervalued players and outperform its budget. The Oakland A's success with data science has revolutionized baseball, with most other baseball teams now integrating similar data-driven strategies into their recruitment processes.

The moneyball story is a very clear example of how data science can give an organization an advantage in a competitive market space. However, from a pure data science perspective perhaps the most important aspect of the moneyball story is that it highlights that sometimes the primary value of data science is the identification of informative attributes. A common belief is that the value of data science is in the models created through the process. However, once we know the important attributes in a domain, it is very easy to create data-driven models. The key to success is getting the right data and finding

The key to success is getting the right data and finding the right attributes.

the right attributes. In *Freakonomics: A Rogue Economist Explores the Hidden Side of Everything*, Steven D. Levitt and Stephen Dubner illustrate the importance of this observation across a wide range of problems. As they put it, the key to understanding modern life is "knowing what to measure and how to measure it" (2009, 14). Using data science, we can uncover the important patterns in a data set, and these patterns can reveal the important attributes in the domain. The reason why data science is used in so many domains is that it doesn't matter what the problem domain is: if the right data are available and the problem can be clearly defined, then data science can help.

Why Now?

A number of factors have contributed to the recent growth of data science. As we have already touched upon, the emergence of big data has been driven by the relative ease with which organizations can gather data. Be it through point-of-sales transaction records, clicks on online platforms, social media posts, apps on smart phones, or myriad other channels, companies can now build much richer profiles of individual customers. Another factor is the commoditization of data storage with economies of scale, making it less expensive than ever before to store data. There has also been tremendous growth in computer power. Graphics cards and graphical processing units (GPUs) were originally developed to do fast graphics rendering for computer games. The distinctive feature of GPUs is that they can carry out fast matrix multiplications. However, matrix multiplications are useful not only for graphics rendering but also for ML. In recent years, GPUs have been adapted and optimized for ML use, which has contributed to large speedups in data processing and model training. User-friendly data science tools have also become available and lowered the barriers to entry into data science. Taken together, these developments mean that it has never been easier to collect, store, and process data.

In the past 10 years there have also been major advances in ML. In particular, deep learning has emerged and has revolutionized how computers can process language and image data. The term *deep learning* describes a family of neural network models with multiple layers of units in the network. Neural networks have been around since the 1940s, but they work best with large, complex data sets and take a great deal of computing resources to train. So the emergence of deep learning is connected with growth in big data and computing power. It is not an exaggeration to describe the impact of deep learning across a range of domains as nothing less than extraordinary.

DeepMind's computer program AlphaGo[7] is an excellent example of how deep learning has transformed a field of research. Go is a board game that originated in China 3,000 years ago. The rules of Go are much simpler than chess; players take turns placing pieces on a board with the goal of capturing their opponent's pieces or surrounding empty territory. However, the simplicity of the rules and the fact that Go uses a larger board means that there are many more possible board configurations in Go then there are in chess. In fact, there are more possible board configurations for Go than there are atoms in the universe. This makes Go much more difficult than chess for computers because of its much larger search space and difficulty in evaluating each of these possible board configurations. The DeepMind team used deep-learning models to enable AlphaGo to evaluate board configurations and to select the next move to make. The result was that AlphaGo became the first computer program to beat a professional Go player, and in March 2016 AlphaGo beat Led Sedol, the 18-time Go world champion, in a match watched by more than 200 million people worldwide. To put the impact of deep learning on Go in context, as recently as 2009 the best Go computer program in the world was rated at the low end of advanced amateur; seven years later AlphaGo beat the world champion. In 2016, an article describing the deep-learning algorithms behind AlphaGo was published in the world's most prestigious academic science journal, *Nature* (Silver, Huang, Maddison, et al. 2016).

Deep learning has also had a massive impact on a range of high-profile consumer technologies. Facebook now uses deep learning for face recognition and to analyze text in order to advertise directly to individuals based on their online conversations. Both Google and Baidu use deep learning for image recognition, captioning and search, and machine translation. Apple's virtual assistant Siri, Amazon's Alexa, Microsoft's Cortana, and Samsung's Bixby use speech recognition based on deep learning. Huawei is currently developing a virtual assistant for the Chinese market, and it, too, will use deep-learning speech recognition. In chapter 4, "Machine Learning 101," we describe neural networks and deep learning in more detail. However, although deep learning is an important technical development, perhaps what is most significant about it in terms of the growth of data science is the increased awareness of the capabilities and benefits of data science and organization buy-in that has resulted from these high-profile success stories.

Myths about Data Science

Data science has many advantages for modern organizations, but there is also a great deal of hype around it, so we should understand what its limitations are. One of the biggest myths is the belief that data science is an autonomous process that we can let loose on our data to find the answers to our problems. In reality, data science requires skilled human oversight throughout the different stages of the process. Humans analysts are needed to frame the problem, to design and prepare the data, to select which ML algorithms are most appropriate, to critically interpret the results of the analysis, and to plan the appropriate action to take based on the insight(s) the analysis has revealed. Without skilled human oversight, a data science project will fail to meet its targets. The best data science outcomes occur when human expertise and computer power work together, as Gordon Linoff and Michael Berry put it: "Data mining lets computers do what they do best—dig through lots of data. This, in turn, lets people do what people do best, which is to set up the problem and understand the results" (2011, 3).

The widespread and growing use of data science means that today the biggest data science challenge for many organizations is locating qualified human analysts and hiring them. Human talent in data science is at a premium, and sourcing this talent is currently the main bottleneck in the adoption of data science. To put this talent shortfall in context, in 2011 a McKinsey Global Institute report projected a shortfall in the United States of between 140,000 and 190,000 people with data science and analytics skills and an even larger shortfall of 1.5 million managers with the ability to understand data science and analytics processes at a level that will enable them to interrogate and interpret the results of data science appropriately (Manyika, Chui, Brown, et al. 2011). Five years on, in their 2016 report, the institute remained convinced that data science has huge untapped value potential across an expanding range of applications but that the talent shortfall will remain, with a predicted shortfall of 250,000 data scientists in the near term (Henke, Bughin, Chui, et al. 2016).

The second big myth of data science is that every data science project needs big data and needs to use deep learning. In general, having more data helps, but having the *right* data is the more important requirement. Data science projects are frequently carried out in organizations that have significantly less resources in terms of data and computing power than Google, Baidu, or Microsoft. Examples indicative of the scale of smaller data science projects include claim prediction in an insurance company that processes around 100 claims a month; student dropout prediction for a university with less than 10,000 students; membership dropout prediction for a union with several thousand members. So an organization doesn't need to be handling terabytes of data or to have massive computing resources at its disposal to benefit from data science.

A third data science myth is that modern data science software is easy to use, and so data science is easy to do. It is true that data science software has become more user-friendly. However, this ease of use can hide the fact that doing data science properly requires both appropriate domain knowledge and the expertise regarding the properties of the data and the assumptions underpinning the different ML algorithms. In fact, it has never been easier to do data science badly. Like everything else in life, if you don't understand what you are doing when you do data science, you are going to make mistakes. The danger with data science is that people can be intimidated by the technology and believe whatever results the software presents to them. They may, however, have unwittingly framed the problem in the wrong way, entered the wrong data, or used analysis techniques with inappropriate assumptions. So the results the software presents are likely to be the answer to the wrong question or to be based on the wrong data or the outcome of the wrong calculation.

The last myth about data science we want to mention here is the belief that data science pays for itself quickly. The truth of this belief depends on the context of the organization. Adopting data science can require significant investment in terms of developing data infrastructure and hiring staff with data science expertise. Furthermore, data science will not give positive results on every project. Sometimes there is no hidden gem of insight in the data, and sometimes the organization is not in a position to act on the insight the analysis has revealed. However, in contexts where there is a well-understood business problem and the appropriate data and human expertise are available, then data science can (often) provide the actionable insight that gives an organization the competitive advantage it needs to succeed.

WHAT ARE DATA,
AND WHAT IS A DATA SET?

As its name suggests, data science is fundamentally dependent on data. In its most basic form, a datum or a piece of information is an abstraction of a real-world entity (person, object, or event). The terms *variable*, *feature*, and *attribute* are often used interchangeably to denote an individual abstraction. Each entity is typically described by a number of attributes. For example, a book might have the following attributes: author, title, topic, genre, publisher, price, date published, word count, number of chapters, number of pages, edition, ISBN, and so on.

A data set consists of the data relating to a collection of entities, with each entity described in terms of a set of attributes. In its most basic form,[1] a data set is organized in an $n * m$ data matrix called the *analytics record*, where n is the number of entities (rows) and m is the number of attributes (columns). In data science, the terms *data set* and *analytics record* are often used interchangeably, with the analytics record being a particular representation of a data set. Table 1 illustrates an analytics record for a data set of classic books. Each row in the table describes one book. The terms *instance*, *example*, *entity*, *object*, *case*, *individual*, and *record* are used in data science literature to refer to a row. So a data set contains a set of instances, and each instance is described by a set of attributes.

The construction of the analytics record is a prerequisite of doing data science. In fact, the majority of the time and effort in data science projects is spent on creating, cleaning, and updating the analytics record. The analytics record is often constructed by merging information from many different sources: data may have to be extracted from multiple databases, data warehouses, or computer files in different formats (e.g., spreadsheets or csv files) or scraped from the web or social media streams.

Table 1   A Data Set of Classic Books

ID	Title	Author	Year	Cover	Edition	Price
1	*Emma*	Austen	1815	Paperback	20th	$5.75
2	*Dracula*	Stoker	1897	Hardback	15th	$12.00
3	*Ivanhoe*	Scott	1820	Hardback	8th	$25.00
4	*Kidnapped*	Stevenson	1886	Paperback	11th	$5.00

Four books are listed in the data set in table 1. Excluding the ID attribute—which is simply a label for each row and hence is not useful for analysis—each book is described using six attributes: title, author, year, cover, edition, and price. We could have included many more attributes for each book, but, as is typical of data science projects, we needed to make a choice when we were designing the data set. In this instance, we were constrained by the size of the page and the number of attributes we could fit onto it. In most data science projects, however, the constraints relate to what attributes we can actually gather and what attributes we believe, based on our domain knowledge, are relevant to the problem we are trying to solve. Including extra attributes in a data set does not come without cost. First, there is the extra time and effort in collecting and quality checking the attribute information for each instance in the data set and integrating these data into the analytics record. Second, including irrelevant or redundant attributes can have a negative effect on the performance of many of the algorithms used to analyze data. Including many attributes in a data set increases the probability that an algorithm will find irrelevant or spurious patterns in the data that appear to be statistically significant only because of the particular sample of instances in the data set. The problem of how to choose the correct attribute(s) is a challenge faced by all data science projects, and sometimes it comes down to an iterative process of trial-and-error experiments where each iteration checks the results achieved using different subsets of attributes.

There are many different types of attributes, and for each attribute type different sorts of analysis are appropriate. So understanding and recognizing different attribute types is a fundamental skill for a data scientist. The standard types are *numeric*, *nominal*, and *ordinal*. Numeric attributes describe measurable quantities that are represented using integer or real values. Numeric attributes can be measured on either an *interval scale* or a *ratio scale*. Interval attributes are measured on a scale with a fixed but arbitrary interval and arbitrary origin—for example, date and time measurements. It is appropriate to apply ordering and subtraction operations to interval attributes, but other arithmetic operations (such as multiplication and division) are not appropriate. Ratio scales are similar to interval scales, but the scale of measurement possesses a true-zero origin. A value of zero indicates that none of the quantity is being measured. A consequence of a ratio scale having a true-zero origin is that we can describe a value on a ratio scale as being a multiple (or ratio) of another value. Temperature is a useful example for distinguishing between interval and ratio scales.[2] A temperature measurement on the Celsius or Fahrenheit scale is an interval measurement because a 0 value on either of these scales does not indicate zero heat. So although we can compute differences between temperatures on these scales and compare these differences, we cannot say that a temperature of 20° Celsius is twice as warm as 10° Celsius. By contrast, a temperature measurement in Kelvins is on a ratio scale because 0 K (absolute zero) is the temperature at which all thermal motion ceases. Other common examples of ratio-scale measurements include money quantities, weight, height, and marks on an exam paper (scale 0–100). In table 1, the "year" attribute is an example of an interval-scale attribute, and the "price" attribute is an example of a ratio-scale attribute.

Nominal (also known as categorical) attributes take values from a finite set. These values are names (hence "nominal") for categories, classes, or states of things. Examples of nominal attributes include marital status (single, married, divorced) and beer type (ale, pale ale, pils, porter, stout, etc.). A binary attribute is a special case of a nominal attribute where the set of possible values is restricted to just two values. For example, we might have the binary attribute "spam," which describes whether an email is spam (true) or not spam (false), or the binary attribute "smoker," which describes whether an individual is a smoker (true) or not (false). Nominal attributes cannot have ordering or arithmetic operations applied to them. Note that a nominal attribute may be sorted alphabetically, but alphabetizing is a distinct operation from ordering. In table 1, "author" and "title" are examples of nominal attributes.

Ordinal attributes are similar to nominal attributes, with the difference that it is possible to apply a rank order over the categories of ordinal attributes. For example, an attribute describing the response to a survey question might take values from the domain "strongly dislike, dislike, neutral, like, and strongly like." There is a natural ordering over these values from "strongly dislike" to "strongly like" (or vice versa depending on the convention being used). However, an important feature of ordinal data is that there is no notion of equal distance between these values. For example, the cognitive distance between "dislike" and "neutral" may be different from the distance between "like" and "strongly like." As a result, it is not appropriate to apply arithmetic operations (such as averaging) on ordinal attributes. In table 1, the "edition" attribute is an example of an ordinal attribute. The distinction between nominal and ordinal data is not always clear-cut. For example, consider an attribute that describes the weather and that can take the values "sunny," "rainy," "overcast." One person might view this attribute as being nominal, with no natural order over the values, whereas another person might argue that the attribute is ordinal, with "overcast" being treated as an intermediate value between "sunny" and "rainy" (Hall, Witten, and Frank 2011).

The data type of an attribute (numeric, ordinal, nominal) affects the methods we can use to analyze and understand the data, including both the basic statistics we can

The data type of an attribute (numeric, ordinal, nominal) affects the methods we can use to analyze and understand the data.

use to describe the distribution of values that an attribute takes and the more complex algorithms we use to identify the patterns of relationships between attributes. At the most basic level of analysis, numeric attributes allow arithmetic operations, and the typical statistical analysis applied to numeric attributes is to measure the central tendency (using the mean value of the attribute) and the dispersion of the attributes values (using the variance or standard deviation statistics). However, it does not make sense to apply arithmetic operations to nominal or ordinal attributes. So the basic analysis of these types of attributes involves counting the number of times each of the values occurs in the data set or calculating the proportion of occurrence of each value or both.

Data are generated through a process of abstraction, so any data are the result of human decisions and choices. For every abstraction, somebody (or some set of people) will have made choices with regard to what to abstract from and what categories or measurements to use in the abstracted representation. The implication is that data are never an objective description of the world. They are instead always partial and biased. As Alfred Korzybski has observed, "A map is not the territory it represents, but, if correct, it has a similar structure to the territory which accounts for its usefulness" (1996, 58).

In other words, the data we use for data science are not a perfect representation of the real-world entities and processes we are trying to understand, but if we are careful in how we design and gather the data that we use, then the results of our analysis will provide useful insights into our real-world problems. The moneyball story given in chapter 1 is a great example of how the determinant of success in many data science projects is figuring out the correct abstractions (attributes) to use for a given domain. Recall that the key to the moneyball story was that the Oakland A's figured out that a player's on-base percentage and slugging percentage are better attributes to use to predict a player's offensive success than traditional baseball statistics such as batting average. Using different attributes to describe players gave the Oakland A's a different and better model of baseball than the other teams had, which enabled it to identify undervalued players and to compete with larger franchises using a smaller budget.

The moneyball story illustrates that the old computer science adage "garbage in, garbage out" is true for data science: if the inputs to a computational process are incorrect, then the outputs from the process will be incorrect. Indeed, two characteristics of data science cannot be overemphasized: (a) for data science to be successful, we need to pay a great deal of attention to how we create our data (in terms of both the choices we make in designing the data abstractions and the quality of the data captured by our abstraction processes), and (b) we also need to "sense check" the results of a data science process—that is, we need to understand that just because the computer identifies a pattern in the data this doesn't mean that it is identifying a real insight in the processes we are trying to analyze; the pattern may simply be based on the biases in our data design and capture.

Perspectives on Data

Other than type of data (numeric, nominal, and ordinal), a number of other useful distinctions can be made regarding data. One such distinction is between *structured* and *unstructured* data. Structured data are data that can be stored in a table, and every instance in the table has the same structure (i.e., set of attributes). As an example, consider the demographic data for a population, where each row in the table describes one person and consists of the same set of demographic attributes (name, age, date of birth, address, gender, education level, job status, etc.). Structured data can be easily stored, organized, searched, reordered, and merged with other structured data. It is relatively easy to apply data science to structured data because, by definition, it is already in a format that is suitable for integration into an analytics record. *Unstructured data* are data where each instance in the data set may have its own internal structure, and this structure is not necessarily the same in every instance. For example, imagine a data set of webpages, with each webpage having a structure but this structure differing from one webpage to another. Unstructured data are much more common than structured data. For example, collections of human text (emails, tweets, text messages, posts, novels, etc.) can be considered unstructured data, as can collections of sound, image, music, video, and multimedia files. The variation in the structure between the different elements means that it is difficult to analyze unstructured data in its raw form. We can often extract structured data from unstructured data using techniques from artificial intelligence (such as natural-language processing and ML), digital signal processing, and computer vision. However, implementing and testing these data-transformation processes is expensive and time-consuming and can add significant financial overhead and time delays to a data science project.

Sometimes attributes are *raw* abstractions from an event or object—for example, a person's height, the number of words in an email, the temperature in a room, the time or location of an event. But data can also be *derived* from other pieces of data. Consider the average salary in a company or the variance in the temperature of a room across a period of time. In both of these examples, the resulting data are derived from an original set of data by applying a function to the original raw data (individual salaries or temperature readings). It is frequently the case that the real value of a data science project is the identification

It is frequently the case that the real value of a data science project is the identification of one or more important derived attributes that provide insight into a problem.

of one or more important derived attributes that provide insight into a problem. Imagine we are trying to get a better understanding of obesity within a population, and we are trying to understand the attributes of an individual that identify him as being obese. We would begin by examining the raw attributes of individuals, such as their height and weight, but after studying the problem for some time we might end up designing a more informative derived attribute such as the Body Mass Index (BMI). BMI is the ratio of a person's mass and height. Recognizing that the *interaction* between the raw attributes "mass" and "height" provides more information about obesity then either of these two attributes can when examined independently will help us to identify people in the population who are at risk of obesity. Obviously, BMI is a simple example that we use here to illustrate the importance of derived attributes. But consider situations where the insight into the problem is given through multiple derived attributes, where each attribute involves two (or potentially more) additional attributes. It is in contexts where multiple attributes interact together that data science provides us with real benefits because the algorithms we use can, in some cases, learn the derived attributes from the raw data.

There are generally two terms for gathered *raw data*: *captured data* and *exhaust data* (Kitchin 2014a). *Captured data* are collected through a direct measurement or observation that is designed to gather the data. For example, the primary purpose of surveys and experiments is to gather specific data on a particular topic of interest. By contrast, exhaust data are a by-product of a process whose primary purpose is something other than data capture. For example, the primary purpose of many social media technologies is to enable users to connect with other people. However, for every image shared, blog posted, tweet retweeted, or post liked, a range of exhaust data is generated: who shared, who viewed, what device was used, what time of day, which device was used, how many people viewed/liked/retweeted, and so on. Similarly, the primary purpose of the Amazon website is to enable users to make purchases from the site. However, each purchase generates volumes of exhaust data: what items the user put into her basket, how long she stayed on the site, what other items she viewed, and so on.

One of the most common types of exhaust data is *metadata*—that is, data that describe other data. When Edward Snowden released documents about the US National Security Agency's surveillance program PRISM, he revealed that the agency was collecting a large amount of metadata about people's phone calls. This meant that the agency was not actually recording the content of peoples phone calls (it was not doing wiretapping) but rather collecting the data about the calls, such as when the call was made, who the recipient was, how long the call lasted, and so on (Pomerantz 2015). This type of data gathering may not appear ominous, but the MetaPhone study carried out at Stanford highlighted the types of sensitive insights that phone-call metadata can reveal about an individual (Mayer and Mutchler 2014). The fact that many organizations have very specific purposes makes it relatively easy to infer sensitive information about a person based on his phone calls to these organizations. For example, some of the people in the MetaPhone study made calls to Alcoholics Anonymous, divorce lawyers, and medical clinics specializing in sexually transmitted diseases. Patterns in calls can also be revealing. The pattern analysis from the study showed how patterns of calls reveal potentially very sensitive information:

> Participant A communicated with multiple local neurology groups, a specialty pharmacy, a rare condition management service, and a hotline for a pharmaceutical used solely to treat relapsing multiple sclerosis. ... In a span of three weeks, Participant D contacted a home improvement store, locksmiths, a hydroponics dealer, and a head shop. (Mayer and Mutchler 2014)

Data science has traditionally focused on captured data. However, as the MetaPhone study shows, exhaust data can be used to reveal hidden insight into situations. In recent years, exhaust data have become more and more useful, particularly in the realm of customer engagement, where the linking of different exhaust data sets has the potential to provide a business with a richer profile of individual customers, thereby enabling the business to target its services and marketing to certain customers. In fact, one of the factors driving the growth in data science in business today is the recognition of the value of exhaust data and the potential that data science has to unlock this value for businesses.

Data Accumulates, Wisdom Doesn't!

The goal of data science is to use data to get insight and understanding. The Bible urges us to attain understanding by seeking wisdom: "wisdom is the principal thing, therefore get wisdom and with all thy getting get understanding" (Proverbs 4:7 [King James]). This advice is reasonable, but it does beg the question of how one should go about seeking wisdom. The following lines from T. S. Eliot's poem "Choruses from The Rock" describes a hierarchy of wisdom, knowledge, and information:

> Where is the wisdom we have lost in knowledge?
> Where is the knowledge we have lost in information?
> (Eliot 1934, 96)

Eliot's hierarchy mirrors the standard model of the structural relationships between wisdom, knowledge, information, and data known as the *DIKW pyramid* (see figure 2). In the DIKW pyramid, data precedes information, which precedes knowledge, which precedes wisdom. Although the order of the layers in the hierarchy are generally agreed upon, the distinctions between the layers and the processes required to move from one layer to the next are often contested. Broadly speaking, however,

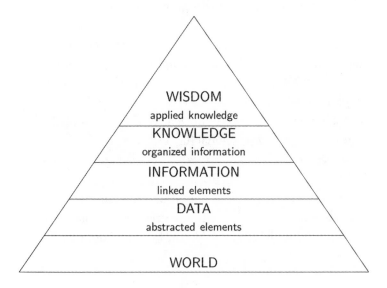

Figure 2   The DIKW pyramid (adapted from Kitchin 2014a).

- Data are created through abstractions or measurements taken from the world.

- Information is data that have been processed, structured, or contextualized so that it is meaningful to humans.

- Knowledge is information that has been interpreted and understood by a human so that she can act on it if required.

- Wisdom is acting on knowledge in an appropriate way.

The activities in the data science process can also be represented using a similar pyramid hierarchy where the width of the pyramid represents the amount of data being processed at each level and where the higher the layer in the pyramid, the more informative the results of the activities are for decision making. Figure 3 illustrates the hierarchy of data science activities from data capture and generation through data preprocessing and aggregation, data understanding and exploration, pattern discovery and model creation using ML, and decision support using data-driven models deployed in the business context.

The CRISP-DM Process

Many people and companies regularly put forward suggestions on the best process to follow to climb the data science pyramid. The most commonly used process is the

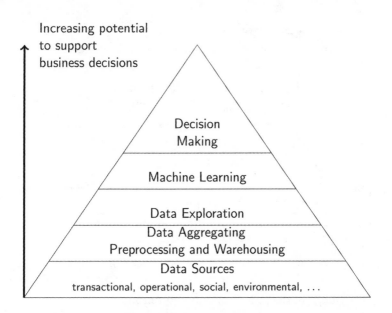

Increasing potential to support business decisions

Decision Making

Machine Learning

Data Exploration

Data Aggregating Preprocessing and Warehousing

Data Sources

transactional, operational, social, environmental, ...

Figure 3   Data science pyramid (adapted from Han, Kamber, and Pei 2011).

Cross Industry Standard Process for Data Mining (CRISP-DM). In fact, the CRISP-DM has regularly been in the number-one spot in various industry surveys for a number of years. The primary advantage of CRISP-DM, the main reason why it is so widely used, is that it is designed to be independent of any software, vendor, or data-analysis technique.

CRISP-DM was originally developed by a consortium of organizations consisting of leading data science vendors, end users, consultancy companies, and researchers. The original CRISP-DM project was sponsored in part by the European Commission under the ESPRIT Program, and the process was first presented at a workshop in 1999. Since then, a number of attempts have been made to update the process, but the original version is still predominantly in use. For many years, there was a dedicated website for CRISP-DM, but in recent years this website is no longer available, and on occasion you might get redirected to the SPSS website by IBM, which was one of the original contributors to the project. The original consortium published a detailed (76-page) but readable step-by-step guide to the process that is freely available online (see Chapman et al. 1999), but the structure and major tasks of the process can be summarized in a few pages.

The CRISP-DM life cycle consists of six stages: *business understanding*, *data understanding*, *data preparation*, *modeling*, *evaluation*, and *deployment*, as shown in figure 4. Data are at the center of all data science activities, and that is why the CRISP-DM diagram has data at its center. The arrows between the stages indicate the typical direction of the process. The process is semistructured, which means that a data scientist doesn't always move through these six stages in a linear fashion. Depending on the outcome of a particular stage, a data scientist may go back to one of

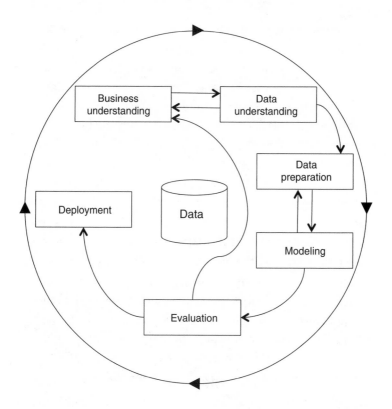

Figure 4 The CRISP-DM life cycle (based on figure 2 in Chapman, Clinton, Kerber, et al. 1999).

the previous stages, redo the current stage, or move on to the next stage.

In the first two stages, business understanding and data understanding, the data scientist is trying to define the goals of the project by understanding the business needs and the data that the business has available to it. In the early stages of a project, a data scientist will often iterate between focusing on the business and exploring what data are available. This iteration typically involves identifying a business problem and then exploring if the appropriate data are available to develop a data-driven solution to the problem. If the data are available, the project can proceed; if not, the data scientist will have to identify an alternative problem to tackle. During this stage of a project, a data scientist will spend a great deal of time in meetings with colleagues in the business-focused departments (e.g., sales, marketing, operations) to understand their problems and with the database administrators to get an understanding of what data are available.

Once the data scientist has clearly defined a business problem and is happy that the appropriate data are available, she moves on to the next phase of the CRISP-DM: data preparation. The focus of the data-preparation stage is the creation of a data set that can be used for the data analysis. In general, creating this data set involves integrating data sources from a number of databases. When an organization has a data warehouse, this data integration can be relatively straightforward. Once a data set has been created, the quality of the data needs to be checked and fixed. Typical data-quality problems include outliers and missing values. Checking the quality of the data is very important because errors in the data can have a serious effect on the performance of the data-analysis algorithms.

The next stage of CRISP-DM is the modeling stage. This is the stage where automatic algorithms are used to extract useful patterns from the data and to create models that encode these patterns. Machine learning is the field of computer science that focuses on the design of these algorithms. In the modeling stage, a data scientist will normally use a number of different ML algorithms to train a number of different models on the data set. A model is trained on a data set by running an ML algorithm on the data set so as to identify useful patterns in the data and to return a model that encodes these patterns. In some cases an ML algorithm works by fitting a template model structure to a data set by setting the parameters of the template to good values for that data set (e.g., fitting a linear regression or neural network model to a data set). In other cases an ML algorithm builds a model in a piecewise fashion (e.g. growing a decision tree one node at a time beginning at the root node of the tree). In most data science projects it is a model generated by an ML algorithm that is ultimately the software that is deployed by an organization to help it solve the problem the data science project is addressing. Each model is trained by a different type of ML algorithm, and each algorithm looks for different types of patterns in the data. At this stage in the project, the data scientist typically doesn't know which patterns are the best ones to look for in the data, so in this context it makes sense to experiment with a number of different algorithms and see which algorithm returns the most accurate models when run on the data set. In chapter 4 we will introduce ML algorithms and models in much more detail and explain how to create a test plan to evaluate model accuracy.

In the majority of data science projects, the initial model test results will uncover problems in the data. These data errors sometimes come to light when the data scientist investigates why the performance of a model is lower than expected or notices that maybe the model's performance is suspiciously good. Or by examining the structure of the models, the data scientist may find that the model is reliant on attributes that she would not expect, and as a result she revisits the data to check that these attributes are correctly encoded. It is thus not uncommon for a project to go through several rounds of these two stages of the process: modeling, data preparation; modeling, data preparation; and so on. For example, Dan Steinberg and his team reported that during one data science project, they rebuilt their data set 10 times over a six-week period, and in week five, having gone through a number of iterations of data cleaning and preparation, they uncovered a major error in the data (Steinberg 2013). If this error had not been identified and fixed, the project would not have succeeded.

The last two stages of the CRISP-DM process, evaluation and deployment, are focused on how the models fit the business and its processes. The tests run during the modeling stage are focused purely on the accuracy of the models for the data set. The evaluation phase involves assessing the models in the broader context defined by the business needs. Does a model meet the business objectives of the process? Is there any business reason why a model is inadequate? At this point in the process, it is also useful for the data scientist to do a general quality-assurance review on the project activities: Was anything missed? Could anything have been done better? Based on the general assessment of the models, the main decision made during the evaluation phase is whether any of the models should be deployed in the business or another iteration of the CRISP-DM process is required to create adequate models. Assuming the evaluation process approves a model or models, the project moves into the final stage of the process: deployment. The deployment phase involves examining how to deploy the selected models into the business environment. This involves planning how to integrate the models into the organization's technical infrastructure and business processes. The best models are the ones that fit smoothly into current practices.

Models that fit current practices have a natural set of users who have a clearly defined problem that the model helps them to solve. Another aspect of deployment is putting a plan in place to periodically review the performance of the model.

The outer circle of the CRISP-DM diagram (figure 4) highlights how the whole process is iterative. The iterative nature of data science projects is perhaps the aspect of these projects that is most often overlooked in discussions of data science. After a project has developed and deployed a model, the model should be regularly reviewed to check that it still fits the business's needs and that it hasn't become obsolete. There are many reasons why a data-driven model can become obsolete: the business's needs might have changed; the process the model emulates and provides insight into might have changed (for example, customer behavior changes, spam email changes, etc.); or the data streams the model uses might have changed (for example, a sensor that feeds information into a model may have been updated, and the new version of the sensor provides slightly different readings, causing the model to be less accurate). The frequency of this review is dependent on how quickly the business ecosystem and the data that the model uses evolve. Constant monitoring is needed to determine the best time to go through the process again. This is what the outer circle of the CRISP-DM process shown in figure 4 represents. For example, depending on the data, the business question, and the domain, you may have go through this iterative process on a yearly, quarterly, monthly, weekly, or even daily basis. Figure 5 gives a summary of the different stages of the data science project process and the major tasks involved in each phase.

A frequent mistake that many inexperienced data scientists make is to focus their efforts on the modeling stage of the CRISP-DM and to rush through the other stages. They may think that the really important deliverable from a project is the model, so the data scientist should devote most of his time to building and finessing the model. However, data science veterans will spend more time on ensuring that the project has a clearly defined focus and that it has the right data. For a data science project to succeed, a data scientist needs to have a clear understanding of the business need that the project is trying to solve. So the business understanding stage of the process is really important. With regard to getting the right data for a project, a survey of data scientists in 2016 found that 79 percent of their time is spent on data preparation. The time spent across the major tasks in the project was distributed as follows: collecting data sets, 19 percent; cleaning and organizing data, 60 percent; building training sets, 3 percent; mining data for patterns, 9 percent; refining algorithms, 4 percent; and performing other tasks, 5 percent (Crowd-Flower 2016). The 79 percent figure for preparation comes from summing the time spent on collecting, cleaning, and

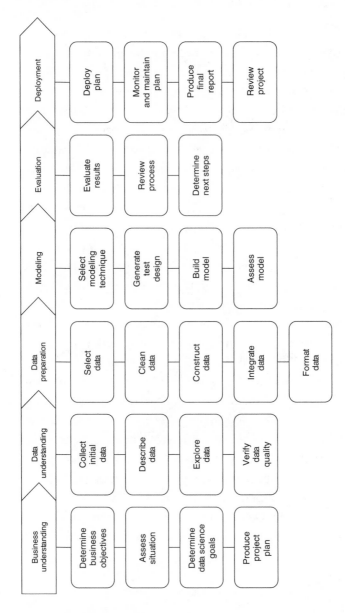

Figure 5 The CRISP-DM stages and tasks (based on figure 3 in Chapman, Clinton, Kerber, et al. 1999).

organizing the data. That around 80 percent of project time is spent on gathering and preparing data has been a consistent finding in industry surveys for a number of years. Sometimes this finding surprises people because they imagine data scientists spend their time building complex models to extract insight from the data. But the simple truth is that no matter how good your data analysis is, it won't identify useful patterns unless it is applied to the right data.

A DATA SCIENCE ECOSYSTEM

The set of technologies used to do data science varies across organizations. The larger the organization or the greater the amount of data being processed or both, the greater the complexity of the technology ecosystem supporting the data science activities. In most cases, this ecosystem contains tools and components from a number of different software suppliers, processing data in many different formats. There is a spectrum of approaches from which an organization can select when developing its own data science ecosystem. At one end of the spectrum, the organization may decide to invest in a commercial integrated tool set. At the other end, it might build up a bespoke ecosystem by integrating a set of open-source tools and languages. In between these two extremes, some software suppliers provide solutions that consist of a mixture of commercial products and open-source products.

However, although the particular mix of tools will vary from one organization to the next, there is a commonality in terms of the components that are present in most data science architectures.

Figure 6 gives a high-level overview of a typical data architecture. This architecture is not just for big-data environments, but for data environments of all sizes. In this diagram, the three main areas consist of *data sources*, where all the data in an organization are generated; *data storage*, where the data are stored and processed; and *applications*, where the data are shared with consumers of these data.

All organizations have applications that generate and capture data about customers, transactions, and operational data on everything to do with how the organization operates. Such data sources and applications include customer management, orders, manufacturing, delivery, invoicing, banking, finance, customer-relationship management (CRM), call center, enterprise resource planning (ERP) applications, and so on. These types of applications are commonly referred to as *online transaction processing* (OLTP) systems. For many data science projects, the data from these applications will be used to form the initial input data set for the ML algorithms. Over time, the volume of data captured by the various applications in the organization grows ever larger and the organization will start to branch out to capture data that was ignored, wasn't

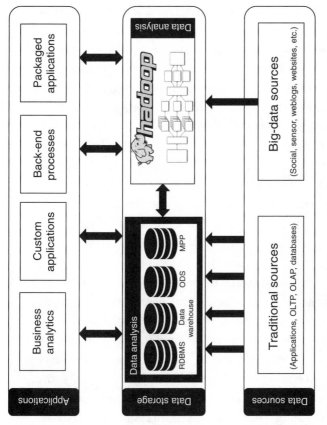

Figure 6 A typical small-data and big-data architecture for data science (inspired by a figure from the Hortonworks newsletter, April 23, 2013, https://hortonworks.com/blog/hadoop -and-the-data-warehouse-when-to-use-which).

captured previously, or wasn't available previously. These newer data are commonly referred to as "big-data sources" because the volume of data that is captured is significantly higher than the organization's main operational applications. Some of the common big-data sources include network traffic, logging data from various applications, sensor data, weblog data, social media data, website data, and so on. In traditional data sources, the data are typically stored in a database. However, because the applications associated with many of the newer big-data sources are not primarily designed to store data long term—for example, with streaming data—the storage formats and structures for this type of data vary from application to application.

As the number of data sources increases, so does the challenge of being able to use these data for analytics and for sharing them across the wider organization. The data-storage layer, shown in figure 6, is typically used to address the data sharing and data analytics across an organization. This layer is divided into two parts. The first part covers the typical data-sharing software used by most organizations. The most popular form of traditional data-integration and storage software is a relational database management system (RDBMS). These traditional systems are often the backbone of the business intelligence (BI) solutions within an organization. A BI solution is a user-friendly decision-support system that provides data aggregating, integration, and reporting as well as analysis functionality. Depending on the maturity level of a BI architecture, it can consist of anything from a basic copy of an operational application to an *operational data store* (ODS) to *massively parallel processing* (MPP) BI database solutions and data warehouses.

Data warehousing is best understood as a process of data aggregation and analysis with the goal of supporting decision making. However, the focus of this process is the creation of a well-designed and centralized data repository, and the term *data warehouse* is sometimes used to denote this type of data repository. In this sense, a data warehouse is a powerful resource for data science. From a data science perspective, one of the major advantages of having a data warehouse in place is a much shorter project time. The key ingredient in any data science process is data, so it is not surprising that in many data science projects the majority of time and effort goes into finding, aggregating, and cleaning the data prior to their analysis. If a data warehouse is available in a company, then the effort and time that go into data preparation on individual data science projects is often significantly reduced. However, it is possible to do data science without a centralized data repository. Constructing a centralized repository of data involves more than simply dumping the data from multiple operational databases into a single database.

Merging data from multiple databases often requires much complex manual work to resolve inconsistencies between the source databases. *Extraction, transformation, and load* (ETL) is the term used to describe the typical processes and tools used to support the mapping, merging, and movement of data between databases. The typical operations carried out in a data warehouse are different from the simple operations normally applied to a standard relational data model database. The term *online analytical processing* (OLAP) is used to describe these operations. OLAP operations are generally focused on generating summaries of historic data and involve aggregating data from multiple sources. For example, we might pose the following OLAP request (expressed here in English for readability): "*Report the sales of all stores by region and by quarter and compare these figures to last year's figures.*" What this example illustrates is that the result of an OLAP request often resembles what you would expect to see as a standard business report. OLAP operations essentially enable users to slice, dice, and pivot the data in the warehouse and get different views of these data. They work on a data representation called a *data cube* that is built on top of the data warehouse. A data cube has a fixed, predefined set of dimensions in which each dimension represents a particular characteristic of the data. The required data-cube dimensions for the example OLAP request given earlier would be *sales by stores*, *sales by region*, and *sales by quarter*. The primary advantage of using a data cube with a fixed set of dimensions is that it speeds up the response time of OLAP operations. Also, because the set of data-cube dimensions is preprogrammed into the OLAP system, the system can provide user-friendly graphical user interfaces for defining OLAP requests. However, the data-cube representation also restricts the types of analysis that can be done using OLAP to the set of queries that can be generated using the predefined dimensions. By comparison, SQL provides a more flexible query interface. Also, although OLAP systems are useful for data exploration and reporting, they don't enable data modeling or the automatic extraction of patterns from the data. Once the data from across an organization has been aggregated and analyzed within the BI system, this analysis can then be used as input to a range of consumers in the applications layer of figure 6.

The second part of the data-storage layer deals with managing the data produced by an organization's big-data sources. In this architecture, the Hadoop platform is used for the storage and analytics of these big data. Hadoop is an open-source framework developed by the Apache Software Foundation that is designed for the processing of big data. It uses distributed storage and processing across clusters of commodity servers. Applying the MapReduce programming model, it speeds up the processing of queries on large data sets. MapReduce implements the *split-apply-combine* strategy: (*a*) a large data set is split up into separate chunks, and each chunk is stored on a different node in the cluster; (b) a query is then applied to all the chunks in parallel; and (c) the result of the query is then calculated by combining the results generated on the different chunks. Over the past couple of years, however, the Hadoop platform is also being used as an extension of an enterprise's data warehouse. Data warehouses originally would store three years of data, but now data warehouses can store more than 10 years of data, and this number keeps increasing. As the amount of data in a data warehouse increases, however, the storage and processing requirements of the database and server also have to increase. This requirement can have a significant cost implication. An alternative is to move some of the older data in a data warehouse for storage into a Hadoop cluster. For example, the data warehouse would store the most recent data, say three years' worth of data, which frequently need to be available for quick analysis and presentation, while the older data and the less frequently used data are stored on Hadoop. Most of the enterprise-level databases have features that connect the data warehouse with Hadoop, allowing a data scientist, using SQL, to query the data in both places as if they all are located in one environment. Her query could involve accessing some data in the data-warehouse database and some of the data in Hadoop. The query processing will be automatically divided into two distinct parts, each running independently, and the results will be automatically combined and integrated before being presented back to the data scientist.

Data analysis is associated with both sections of the data-storage layer in figure 6. Data analysis can occur on the data in each section of the data layer, and the results from data analysis can be shared between each section while additional data analysis is being performed. The data from traditional sources frequently are relatively clean and information dense compared to the data captured from big-data sources. However, the volume and real-time nature of many big-data sources means that the effort involved in preparing and analyzing these big-data sources can be repaid in terms of additional insights not available through the data coming from traditional sources. A variety of data-analysis techniques developed across a number of different fields of research (including natural-language processing, computer vision, and ML) can be used to transform unstructured, low-density, low-value big data into high-density and high-value data. These high-value data can then be integrated with the other high-value data from traditional sources for further data analysis. The description given in this chapter and illustrated in figure 6 is the typical architecture of the data science ecosystem. It is suitable for most organizations, both small and large. However, as an organization scales in size, so too will the complexity of its data science ecosystem. For example, smaller-scale organizations may not require the Hadoop component, but for very large organizations the Hadoop component will become very important.

Moving the Algorithms to the Data

The traditional approach to data analysis involves the extraction of data from various databases, integrating the data, cleaning the data, subsetting the data, and building predictive models. Once the prediction models have been created they can be applied to the new data. Recall from chapter 1 that a prediction model predicts the missing value of an attribute: a spam filter is a prediction model that predicts whether the classification attribute of an email should have the value of "spam" or not. Applying the predictive models to the instances in new data to generate the missing values is known as "scoring the data." Then the final results, after scoring new data, may be loaded back into a database so that these new data can be used as part of some workflow, reporting dashboard, or some other company assessment practice. Figure 7 illustrates that much of the data processing involved in data preparation and analysis is located on a server that is separate from the databases and the data warehouse. Therefore, a significant amount of time can be spent just moving the data out of the database and moving the results back into the database.

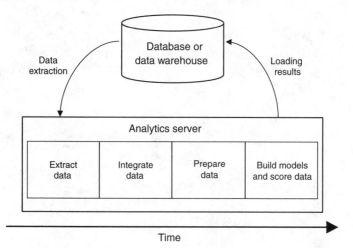

Figure 7 The traditional process for building predictive models and scoring data.

An experiment run at the Dublin Institute of Technology on building a linear-regression model supplies an example of the time involved in each part of the process. Approximately 70 to 80 percent of the time is taken with extracting and preparing the data; the remaining time is spent on building the models. For scoring data, approximately 90 percent of the time is taken with extracting the data and saving the scored data set back into the database; only 10 percent of the time is spent on actually scoring. These results are based on data sets consisting of anywhere from 50,000 records up to 1.5 million records.

A significant amount of
time can be spent just
moving the data out
of the database and
moving the results back
into the database.

Most enterprise database vendors have recognized the time savings that would be available if time did not have to be spent on moving data and have responded to this problem by incorporating data-analysis functionality and ML algorithms into their database engines. The following sections explore how ML algorithms have been integrated into modern databases, how data storage works in the big-data world of Hadoop, and how using a combination of these two approaches allows organizations to easily work with all their data using SQL as a common language for accessing, analyzing, and performing ML and predictive analytics in real time.

The Traditional Database or the Modern Traditional Database

Database vendors continuously invest in developing the scalability, performance, security, and functionality of their databases. Modern databases are far more advanced than traditional relational databases. They can store and query data in variety of different formats. In addition to the traditional relational formats, it is also possible to define object types, store documents, and store and query JSON objects, spatial data, and so on. Most modern databases also come with a large number of statistical functions, so that some have an equivalent number of statistical functions as most statistical applications. For example, the Oracle Database comes with more than

300 different statistical functions and the SQL language built into it. These statistical functions cover the majority of the statistical analyses needed by data science projects and include most if not all the statistical functions available in other tools and languages, such as R. Using the statistical functionality that is available in the databases in an organization may allow data analytics to be performed in a more efficient and scalable manner using SQL. Furthermore, most leading database vendors (including Oracle, Microsoft, IBM, and EnterpriseDB) have integrated many ML algorithms into their databases, and these algorithms can be run using SQL. ML that is built into the database engine and is accessible using SQL is known as *in-database machine learning*. In-database ML can lead to quicker development of models and quicker deployment of models and results to applications and analytic dashboards. The idea behind the in-database ML algorithms is captured in the following directive: "*Move the algorithms to the data instead of the data to the algorithms.*"

The main advantages of using the in-database ML algorithms are:

• **No data movement**. Some data science products require the data to be exported from the database and converted to a specialized format for input to the ML algorithms. With in-database ML, no data movement or conversion is needed. This makes the entire process less complex, less time-consuming, and less error prone.

• **Faster performance**. With analytical operations performed in the database and with no data movement, it is possible to utilize the computing capabilities of the database server, delivering performance up to 100 times faster than the traditional approach. Most database servers have high specifications, with many central processing units (CPUs) and efficient memory management to process data sets containing more than one billion records.

• **High security**. The database provides controlled and auditable access to the data in the database, accelerating the data scientist's productivity while maintaining data security. Also, in-database ML avoids the physical security risks inherent in extracting and downloading data to alternative analytics servers. The traditional process, in contrast, results in the creation of many copies (and potentially different versions) of data sets in separate silos across the organization.

• **Scalability**. A database can easily scale the analytics as the data volume increases if the ML algorithms are brought into the database. The database software is designed to manage large volumes of data efficiently, utilizing the multiple CPUs and memory on the server to allow the ML algorithms to run in parallel. Databases are also very efficient at processing large data sets that do not fit easily into memory. Databases have more than 40 years of development work behind them to ensure that they can process datasets quickly.

• **Real-time deployment and environments**. The models that are developed using the in-database ML algorithms can be immediately deployed and used in real-time environments. This allows the integration of the models into everyday applications, providing real-time predictions to end users and customers.

• **Production deployment**. Models developed using standalone ML software may have to be recoded into other programming languages before they can be deployed into enterprise applications. This is not the case with in-database ML. SQL is the language of the database; it can be used and called by any programming language and data science tool. It is then a simple task to incorporate the in-database models into production applications.

Many organizations are exploiting the benefits of in-database ML. They range from small and medium organizations to large, big-data-type organizations. Some examples of organizations that use in-database ML technologies are:

• Fiserv, an American provider of financial services and fraud detection and analysis. Fiserv migrated from using multiple vendors for data storage and ML to using just the ML capabilities in its database. By using in-database ML, the time used for creating/updating and deploying a fraud-detection model went from nearly a week to just a few hours.

• 84.51˚ (formally Dunnhumby USA), a customer science company. 84.51˚ used many different analytic products to create its various customer models. It typically would spend more than 318 hours each month moving data from its database to its ML tools and back again, plus an additional 67 hours a month to create models. When it switched to using the ML algorithms in its database, there was no more need for data movement. The data stayed in the database. The company immediately saved more than 318 hours of time per month. Because it was using its database as a compute engine, it was able to scale its analytics, and the time taken to generate or update its ML models went from more than 67 hours to one hour per month. This gave the company a saving of sixteen days each month. It is now able to get significantly quicker results and can now provide its customers with results much sooner after they have made a purchase.

• Wargaming, the creators of *World of Tanks* and many other games. Wargaming uses in-database ML to model and predict how to interact with their more than 120 million customers.

Big Data Infrastructure

Although the traditional (modern) database is incredibly efficient at processing transactional data, in the age of big data new infrastructure is required to manage all the other forms of data and for longer-term storage of the data. The modern traditional database can cope with data volumes up to a few petabytes, but for this scale of data, traditional database solutions may become prohibitively expensive. This cost issue is commonly referred to as *vertical scaling*. In the traditional data paradigm, the more data an organization has to store and process within a reasonable amount of time, the larger the database server required and in turn the greater the cost for server configuration and database licensing. Organizations may be able to ingest and query one billion records on a daily/weekly bases using traditional databases, but for this scale of processing they may need to invest more than $100,000 just purchasing the required hardware.

Hadoop is an open-source platform developed and released by the Apache Software Foundation. It is a well-proven platform for ingesting and storing large volumes of data in an efficient manner and can be much less expensive than the traditional database approach. In Hadoop, the data are divided up and partitioned in a variety of ways, and these partitions or portions of data are spread across the nodes of the Hadoop cluster. The various analytic tools that work with Hadoop process the data that reside on each of the nodes (in some instances these data can be memory resident), thus allowing for speedy processing of the data because the analytics is performed in parallel across the nodes. No data extraction or ETL process is needed. The data are analyzed where they are stored.

Although Hadoop is the best known big-data processing framework, it is by no means the only one. Other big-data processing frameworks include Storm, Spark, and Flink. All of these frameworks are part of the Apache software foundation projects. The difference between these frameworks lies in the fact that Hadoop is primarily designed for batch processing of data. Batch processing is appropriate where the dataset is static during the processing and where the results of the processing are not required immediately (or at least are not particularly time sensitive). The Storm framework is designed for processing streams of data. In stream processing each element in the stream is processed as it enters the system, and consequently the processing operations are defined to work on each individual element in the stream rather than on the entire data set. For example, where a batch process might return an average over a data set of values, a stream process will return an individual label or value for each element in the stream (such as calculating a sentiment score for each tweet in a Twitter stream). Storm is designed for real-time processing of data and according to the Storm website,[1] it has been benchmarked at processing over a million tuples per second per node. Spark and Flink are both hybrid (batch and stream) processing frameworks. Spark is a fundamentally a batch processing framework, similar to Hadoop, but also has some stream processing capabilities whereas Flink is a stream processing framework that can also be used for batch processing. Although these big-data processing frameworks provide data scientists with a choice of tools to meet the specific big-data requirements of their project using these frameworks can have the drawback that the modern data scientist now has to analyze data in two different locations, in the traditional modern databases and in the big-data storage. The next section looks at how this particular issue is being addressed.

The Hybrid Database World
If an organization does not have data of the size and scale that require a Hadoop solution, then it will require only traditional database software to manage its data. However, some of the literature argues that the data-storage and processing tools available in the Hadoop world will replace the more traditional databases. It is very difficult to see this happening, and more recently there has been much discussion about having a more balanced approach to managing data in what is called the "hybrid database world." The hybrid database world is where traditional databases and the Hadoop world coexist.

In the hybrid database world, the organization's databases and Hadoop-stored data are connected and work

The hybrid database automatically balances the location of the data based on the frequency of access and the type of data science being performed.

together, allowing the efficient processing, sharing, and analysis of the data. Figure 8 shows a traditional data warehouse, but instead of all the data being stored in the database or the data warehouse, the majority of the data is moved to Hadoop. A connection is created between the database and Hadoop, which allows the data scientist to query the data as if they all are in one location. The data scientist does not need to query the portion of data that is in the database warehouse and then in a separate step query the portion that is stored in Hadoop. He can query the data as he always has done, and the solution will identify what parts of the query need to be run in each location. The results of the query arrived at in each location will be merged together and presented to him. Similarly, as the data warehouse grows, some the older data will not be queried as frequently. The hybrid database solution automatically moves the less frequently used data to the Hadoop environment and the more frequently used data to the warehouse. The hybrid database automatically balances the location of the data based on the frequency of access and the type of data science being performed.

One of the advantages of this hybrid solution is that the data scientist still uses SQL to query the data. He does not have to learn another data-query language or have to use a variety of different tools. Based on current trends, the main database vendors, data-integration solution vendors, and all cloud data-storage vendors will have solutions similar to this hybrid one in the near future.

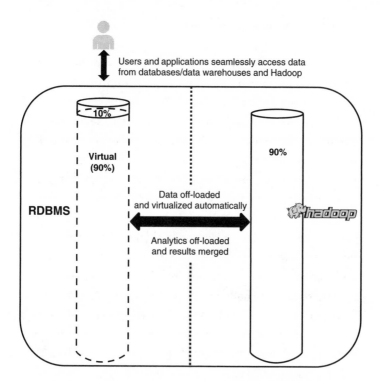

Figure 8 Databases, data warehousing, and Hadoop working together (inspired by a figure in the Gluent data platform white paper, 2017, https://gluent.com/wp-content/uploads/2017/09/Gluent-Overview.pdf).

Data Preparation and Integration

Data integration involves taking the data from different data sources and merging them to give a unified view of the data from across the organization. A good example of such integration occurs with medical records. Ideally, every person would have one health record, and every hospital, medical facility, and general practice would use the same patient identifier or same units of measures, the same grading system, and so on. Unfortunately, nearly every hospital has its own independent patient-management system, as does each of the medical labs within the hospital. Think of the challenges in finding a patient's record and assigning the correct results to the correct patient. And these are the challenges faced by just one hospital. In scenarios where multiple hospitals share patient data, the problem of integration becomes significant. It is because of these kind of challenges that the first three CRISP-DM stages take up to 70 to 80 percent of the total data science project time, with the majority of this time being allocated to data integration.

Integrating data from multiple data sources is difficult even when the data are structured. However, when some of the newer big-data sources are involved, where semi- or unstructured data are the norm, then the cost of integrating the data and managing the architecture can become significant. An illustrative example of the challenges of data integration is customer data. Customer data can reside in many different applications (and the applications' corresponding databases). Each application will contain a slightly different piece of customer data. For example, the internal data sources might contain the customer credit rating, customer sales, payments, call-center contact information, and so on. Additional data about the customer may also be available from external data sources. In this context, creating an integrated view of a customer requires the data from each of these sources to be extracted and integrated.

The typical data-integration process will involve a number of different stages, consisting of extracting, cleaning, standardizing, transforming, and finally integrating to create a single unified version of the data. Extracting data from multiple data sources can be challenging because many data sources can be accessed only by using an interface particular to that data source. As a consequence, data scientists need to have a broad skill set to be able to interact with each of the data sources in order to obtain the data.

Once data have been extracted from a data source, the quality of the data needs to be checked. Data cleaning is a process that detects, cleans, or removes corrupt or inaccurate data from the extracted data. For example, customer address information may have to be cleaned in order to convert it into a standardized format. In addition, there may be duplicate data in the data sources, in which case it is necessary to identify the correct customer record that should be used and to remove all the other records from the data sets. It is important to ensure that the values used in a data set are consistent. For example, one source application might use numeric values to represent a customer credit rating, but another might have a mixture of numeric and character values. In such a scenario, a decision regarding what value to use is needed, and then the other representations should be mapped into the standardized representation. For example, imagine one of the attributes in the data set is a customer's shoe size. Customers can buy shoes from various regions around the world, but the numbering system used for shoe sizes in Europe, the United States, the United Kingdom, and other countries are slightly different. Prior to doing data analysis and modeling, these data values need to be standardized.

Data transformation involves the changing or combining of the data from one value to another. A wide variety of techniques can be used during this step and include data smoothing, binning, and normalization as well as writing custom code to perform a particular transformation. A common example of data transformation is with processing a customer's age. In many data science tasks, precisely distinguishing between customer ages is not particularly helpful. The difference between a 42-year-old customer and a 43-year-old customer is generally not significant, although differentiating between a 42-year-old customer and a 52-year-old customer may be informative. As a consequence, a customer's age is often transformed from a raw age into a general age range. This process of converting ages into age ranges is an example of a data-transformation technique called *binning*. Although binning is relatively straightforward from a technical perspective, the challenge here is to identify the most appropriate range thresholds to apply during binning. Applying the wrong thresholds may obscure important distinctions in the data. Finding appropriate thresholds, however, may require domain specific knowledge or a process of trial-and-error experimentation.

The final step in data integration involves creating the data that are used as input to the ML algorithms. This data is known as the *analytics base table*.

Creating the Analytics Base Table

The most important step in creating the analytics base table is the selection of the attributes that will be included in the analysis. The selection is based on domain knowledge and on an analysis of the relationships between attributes. Consider, for example, a scenario where the analysis is focused on customers of a service. In this scenario, some of the frequently used domain concepts that will inform the design and selection of attributes include customer contract details, demographics, usage, changes in usage, special usage, life-cycle phase, network links, and so on. Furthermore, attributes that are found to have a high correlation with other attributes are likely to be redundant, and so one of the correlated attributes should be excluded. Removing redundant features can result in simpler models which are easier to understand, and also reduces the likelihood of an ML algorithm returning a model that is fitted to spurious patterns in the data. The set of attributes selected for inclusion define what is known as the *analytics record*. An analytics record typically includes both raw and derived attributes. Each instance in the analytics base table is represented by one analytics record, so the set of attributes included in the analytics record defines the representation of the instances the analysis will be carried out on.

After the analytics record has been designed, a set of records needs to extracted and aggregated to create a data set for analysis. When these records have been created and stored—for example, in a database—this data set is commonly referred to as the *analytics base table*. The analytics base table is the data set that is used as input to the ML algorithms. The next chapter introduces the field of ML and describes some of the most popular ML algorithms used in data science.

MACHINE LEARNING 101

Data science is best understood as a partnership between a data scientist and a computer. In chapter 2, we described the process the data scientist follows: the CRISP-DM life cycle. CRISP-DM defines a sequence of decisions the data scientist has to make and the activities he should engage in to inform and implement these decisions. In CRISP-DM, the major tasks for a data scientist are to define the problem, design the data set, prepare the data, decide on the type of data analysis to apply, and evaluate and interpret the results of the data analysis. What the computer brings to this partnership is the ability to process data and search for patterns in the data. Machine learning is the field of study that develops the algorithms that the computers follow in order to identify and extract patterns from data. ML algorithms and techniques are applied primarily during the modeling stage of CRISP-DM. ML involves a two-step process.

First, an ML algorithm is applied to a data set to identify useful patterns in the data. These patterns can be represented in a number of different ways. We describe some popular representations later in this chapter, but they include decision trees, regression models, and neural networks. These representations of patterns are known as "models," which is why this stage of the CRISP-DM life cycle is known at the "modeling stage." Simply put, ML algorithms create models from data, and each algorithm is designed to create models using a particular representation (neural network or decision tree or other).

Second, once a model has been created, it is used for analysis. In some cases, the structure of the model is what is important. A model structure can reveal what the important attributes are in a domain. For example, in a medical domain we might apply an ML algorithm to a data set of stroke patients and use the structure of the model to identify the factors that have a strong association with stroke. In other cases, a model is used to label or classify new examples. For instance, the primary purpose of a spam-filter model is to label new emails as either spam or not spam rather than to reveal the defining attributes of spam email.

Supervised versus Unsupervised Learning

The majority of ML algorithms can be classified as either *supervised learning* or *unsupervised learning*. The goal of supervised learning is to learn a function that maps from the values of the attributes describing an instance to the value of another attribute, known as the *target attribute*, of that instance. For example, when supervised learning is used to train a spam filter, the algorithm attempts to learn a function that maps from the attributes describing an email to a value (spam/not spam) for the target attribute; the function the algorithm learns is the spam-filter model returned by the algorithm. So in this context the pattern that the algorithm is looking for in the data is the function that maps from the values of the input attributes to the values of the target attribute, and the model that the algorithm returns is a computer program that implements this function. Supervised learning works by searching through lots of different functions to find the function that best maps between the inputs and output. However, for any data set of reasonable complexity there are so many combinations of inputs and possible mappings to outputs that an algorithm cannot try all possible functions. As a consequence, each ML algorithm is designed to look at or prefer certain types of functions during its search. These preferences are known as the algorithm's *learning bias*. The real challenge in using ML is to find the algorithm whose learning bias is the best match for a particular data set. Generally, this task involves experiments with a number of different algorithms to find out which one works best on that data set.

Supervised learning is "supervised" because each of the instances in the data set lists both the input values and the output (target) value for each instance. So the learning algorithm can guide its search for the best function by checking how each function it tries matches with the data set, and at the same time the data set acts as a supervisor for the learning process by providing feedback. Obviously, for supervised learning to take place, each instance in the data set must be labeled with the value of the target attribute. Often, however, the reason a target attribute is interesting is that it is not easy to directly measure, and therefore it is not possible to easily create a data set of labeled instances. In such scenarios, a great deal of time and effort is required to create a data set with the target values before a model can be trained using supervised learning.

In unsupervised learning, there is no target attribute. As a consequence, unsupervised-learning algorithms can be used without investing the time and effort in labeling the instances of the data set with a target attribute. However, not having a target attribute also means that learning becomes more difficult: instead of the specific problem of searching for a mapping from inputs to output that

The real challenge in using ML is to find the algorithm whose learning bias is the best match for a particular data set.

matches the data, the algorithm has the more general task of looking for regularities in the data. The most common type of unsupervised learning is *cluster analysis*, where the algorithm looks for clusters of instances that are more similar to each other than they are to other instances in the data. These clustering algorithms often begin by guessing a set of clusters and then iteratively updating the clusters (dropping instances from one cluster and adding them to another) so as to increase both the within-cluster similarity and the diversity across clusters.

A challenge for clustering is figuring out how to measure similarity. If all the attributes in a data set are numeric and have similar ranges, then it probably makes sense just to calculate the Euclidean distance (better known as the *straight-line distance*) between the instances (or rows). Rows that are close together in the Euclidean space are then treated as similar. A number of factors, however, can make the calculation of similarity between rows complex. In some data sets, different numeric attributes have different ranges, with the result that a variation in row values in one attribute may not be as significant as a variation of a similar magnitude in another attribute. In these cases, the attributes should be normalized so that they all have the same range. Another complicating factor in calculating similarity is that things can be deemed similar in many different ways. Some attributes are sometimes more important than other attributes, so it might make sense to weight some attributes in the distance calculations, or it may be that the data set includes nonnumeric data. These more complex scenarios may require the design of bespoke similarity metrics for the clustering algorithm to use.

Unsupervised learning can be illustrated with a concrete example. Imagine we are interested in analyzing the causes of Type 2 diabetes in white American adult males. We would begin by constructing a data set, with each row representing one person and each column representing an attribute that we believe are relevant for the study. For this example, we will include the following attributes: an individual's height in meters and weight in kilos, the number of minutes he exercises per week, his shoe size, and the likelihood that he will develop diabetes expressed as a percentage based on a number of clinical tests and lifestyle surveys. Table 2 illustrates a snippet from this data

Table 2 Diabetes Study Data Set

ID	Height (meters)	Weight (kilograms)	Shoe Size	Exercise (minutes per week)	Diabetes (% likelihood)
1	1.70	70	5	130	0.05
2	1.77	88	9	80	0.11
3	1.85	112	11	0	0.18

...

set. Obviously, other attributes could be included—for example, a person's age—and some attributes could be removed—for example, shoe size, which wouldn't be particularly relevant in determining whether someone will develop diabetes. As we discussed in chapter 2, the choice of which attributes to include and exclude from a data set is a key task in data science, but for the purposes of this discussion we will work with the data set as is.

An unsupervised clustering algorithm will look for groups of rows that are more similar to each other than they are to the other rows in the data. Each of these groups of similar rows defines a cluster of similar instances. For instance, an algorithm can identify causes of a disease or disease comorbidities (diseases that occur together) by looking for attribute values that are relatively frequent within a cluster. The simple idea of looking for clusters of similar rows is very powerful and has applications across many areas of life. Another application of clustering rows is making product recommendations to customers. If a customer liked a book, song, or movie, then he may enjoy another book, song, or movie from the same cluster.

Learning Prediction Models

Prediction is the task of estimating the value of a target attribute for a given instance based on the values of other attributes (or input attributes) for that instance. It is the problem that supervised ML algorithms solve: they generate prediction models. The spam-filter example we used to illustrate supervised learning is also applicable here: we use supervised learning to train a spam-filter model, and the spam-filter model is a prediction model. The typical use case of a prediction model is to estimate the target attribute for new instances that are not in the training data set. Continuing our spam example, we train our spam filter (prediction model) on a data set of old emails and then use the model to predict whether new emails are spam or not spam. Prediction problems are possibly the most popular type of problem that ML is used for, so the rest of this chapter focuses on prediction as the case study for introducing ML. We begin our introduction to prediction models with a concept fundamental to prediction: *correlation analysis*. Then we explain how supervised ML algorithms work to create different types of popular prediction models, including linear-regression models, neural network models, and decision trees.

Correlations Are Not Causations, but Some Are Useful

A *correlation* describes the strength of association between two attributes.[1] In a general sense, a correlation can describe any type of association between two attributes. The term *correlation* also has a specific statistical meaning, in which it is often used as shorthand for "Pearson correlation." A Pearson correlation measures the strength of a linear relationship between two numeric attributes. It ranges in value from −1 to +1. The letter r is used to denote the Pearson value or coefficient between two attributes. A coefficient of $r = 0$ indicates that the two attributes are not correlated. A coefficient of $r = +1$ indicates that the two attributes have a perfect positive correlation, meaning that every change in one attribute is accompanied by an equivalent change in the other attribute in the same direction. A coefficient of $r = −1$ indicates that the two attributes have a perfect negative correlation, meaning that every change in one attribute is accompanied by the opposite change in the other attribute. The general guidelines for interpreting Pearson coefficients are that a value of $r \approx \pm 0.7$ indicates a strong linear relationship between the attributes, $r \approx \pm 0.5$ indicates a moderate linear relationship, $r \approx \pm 0.3$ indicates a weak relationship, and $r \approx 0$ indicates no relationship between the attributes.

In the case of the diabetes study, from our knowledge of how humans are physically made we would expect that there will be relationships between some of the attributes listed in table 2. For example, it is generally the case that the taller someone is, the larger her shoe size is. We would also expect that the more someone exercises, the lighter she will be, with the caveat that a tall person is likely to be heavier than a shorter person who exercises the same amount. We would also expect that there will be no obvious relationship between someone's shoe size and the amount she exercises. Figure 9 presents three scatterplots that illustrate how these intuitions are reflected in the data. The scatterplot at the top shows how the data spread out if the plotting is based on shoe size and height. There is a clear pattern in this scatterplot: the data go from the bottom-left corner to the top-right corner, indicating the relationship that as people get taller (or as we move to the right on the x axis), they also tend to wear larger shoes (we move up on the y axis). A pattern of data generally going from bottom left to top right in a scatterplot is indicative of a positive correlation between the two attributes. If we compute the Pearson correlation between shoe size and height, the correlation coefficient is $r = 0.898$, indicating a strong positive correlation between this pair of attributes. The middle scatterplot shows how the data spread out when we plot weight and exercise. Here the general pattern is in the opposite direction, from top left to bottom right, indicating a negative correlation: the more people exercise, the lighter they are. The Pearson correlation coefficient for this pair of attributes is $r = -0.710$, indicating a strong negative correlation. The final scatterplot, at the bottom, plots exercise and shoe size. The data are relatively randomly distributed in this plot, and the Pearson correlation coefficient for this pair of attributes is $r = -0.272$, indicating no real correlation.

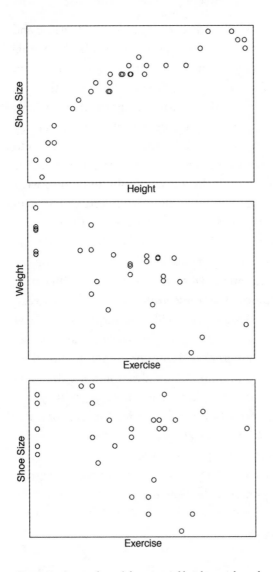

Figure 9 Scatterplots of shoe size and height, weight and exercise, and shoe size and exercise.

The fact that the definition of a statistical Pearson correlation is between two attributes might appear to limit the application of statistical correlation to data analysis to just pairs of attributes. Fortunately, however, we can circumvent this problem by using functions over sets of attributes. In chapter 2, we introduced BMI as a function of a person's weight and height. Specifically, it is the ratio of his weight (in kilograms) divided by his height (in meters) squared. BMI was invented in the nineteenth century by a Belgian mathematician, Adolphe Quetelet, and is used to categorize individuals as underweight, normal weight, overweight, or obese. The ratio of weight and height is used because BMI is designed to have a similar value for people who are in the same category (underweight, normal weight, overweight, or obese) irrespective of their height. We know that weight and height are positively correlated (generally, the taller someone is, the heavier he is), so by dividing weight by height, we control for the effect of height on weight. We divide by the square of the height because people get wider as they get taller, so squaring the height is an attempt to account for a person's total volume in the calculation. Two aspects of BMI are interesting for our discussion about correlation between multiple attributes. First, BMI is a function that takes a number of attributes as input and maps them to a new value. In effect, this mapping creates a new derived (as opposed to raw) attribute in the data. Second, because a person's BMI is a single numeric value, we can calculate the correlation between it and other attributes.

In our case study of the causes of Type 2 diabetes in white American adult males, we are interested in identifying if any of the attributes have a strong correlation with the target attribute describing a person's likelihood of developing diabetes. Figure 10 presents three more scatterplots, each plotting the correlation between the target attribute (diabetes) and another attribute: height, weight, and BMI. In the scatterplot of height and diabetes, there doesn't appear to be any particular pattern in the data indicating that there is no real correlation between these two attributes (the Pearson coefficient is $r = -0.277$). The middle scatterplot shows the distribution of the data plotted using weight and diabetes. The spread of the data indicates a positive correlation between these two attributes: the more someone weighs, the more likely she is to develop diabetes (the Pearson coefficient is $r = 0.655$). The bottom scatterplot shows the data set plotted using BMI and diabetes. The pattern in this scatterplot is similar to the middle scatterplot: the data spread from bottom left to top right, indicating a positive correlation. In this scatterplot, however, the instances are more tightly packed together, indicating that the correlation between BMI and diabetes is stronger than the correlation between weight and diabetes. In fact, the Pearson coefficient for diabetes and BMI for this data set is $r = 0.877$.

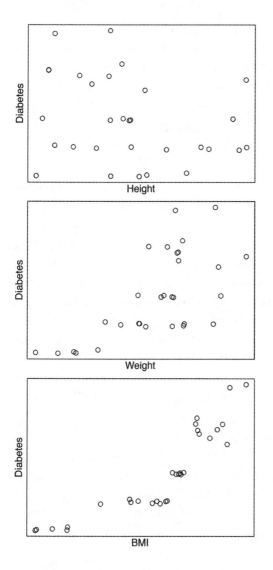

Figure 10   Scatterplots of the likelihood of diabetes with respect to height, weight, and BMI.

The BMI example illustrates that it is possible to create a new derived attribute by defining a function that takes multiple attributes as input. It also shows that it is possible to calculate a Pearson correlation between this derived attribute and another attribute in the data set. Furthermore, a derived attribute can actually have a higher correlation with a target attribute than any of the attributes used to generate the derived attribute have with the target. One way of understanding why BMI has a higher correlation with the diabetes attribute compared to the correlation for either height or weight is that the likelihood of someone developing diabetes is dependent on the interaction between height and weight, and the BMI attribute models this interaction appropriately for diabetes. Clinicians are interested in people's BMI because it gives them more information about the likelihood of someone developing Type 2 diabetes than either just the person's height or just his weight does independently.

We have already noted that attribute selection is a key task in data science. So is attribute design. Designing a derived attribute that has a strong correlation with an attribute we are interested in is often where the real value of data science is found. Once you know the correct attributes to use to represent the data, you are able to build accurate models relatively quickly. Uncovering and designing the right attributes is the difficult part. In the case of BMI, a human designed this derived attribute in the nineteenth century. However, ML algorithms can learn interactions between attributes and create useful derived attributes by searching through different combinations of attributes and checking the correlation between these combinations and the target attribute. This is why ML is useful in contexts where many weak interacting attributes contribute to the process we are trying to understand.

Identifying an attribute (raw or derived) that has a high correlation with a target attribute is useful because the correlated attribute may give us insight into the process that causes the phenomenon the target attribute represents: the fact that BMI is strongly correlated with the likelihood of a person's developing diabetes indicates that it is not weight by itself that contributes to a person's developing diabetes but whether that person is overweight. Also, if an input attribute is highly correlated with a target attribute, it is likely to be a useful input into the prediction model. Similar to correlation analysis, prediction involves analyzing the relationships between attributes. In order to be able to map from the values of a set of input attributes to a target attribute, there must be a correlation between the input attributes (or some derived function over them) and the target attribute. If this correlation does not exist (or cannot be found by the algorithm), then the input attributes are irrelevant for the prediction problem, and the best a model can do is to ignore those inputs and always predict the central tendency of that target[2] in the data set.

Conversely, if a strong correlation does exist between input attributes and the target, then it is likely that an ML algorithm will be able to generate a very accurate prediction model.

Linear Regression

When a data set is composed of numeric attributes, then prediction models based on regression are frequently used. *Regression analysis* estimates the expected (or average) value of a numeric target attribute when all the input attributes are fixed. The first step in a regression analysis is to hypothesize the structure of the relationship between the input attributes and the target. Then a parameterized mathematical model of the hypothesized relationship is defined. This parameterized model is called a *regression function*. You can think of a regression function as a machine that converts inputs to an output value and of the parameters as the settings that control the behavior of a machine. A regression function may have multiple parameters, and the focus of regression analysis is to find the correct settings for these parameters.

It is possible to hypothesize and model many different types of relationships between attributes using regression analysis. In principle, the only constraint on the structure of the relationship that can be modeled is the ability to define the appropriate regression function. In some domains, there may be strong theoretical reasons to assert a particular type of relationship, but in the absence of this type of domain theory it is good practice to begin by assuming the simplest form of relationship—namely, a linear relationship—and then, if need be, progress to model more complex relationships. One reason for starting with a linear relationship is that linear-regression functions are relatively easy to interpret. The other reason is the commonsense notion that keeping things as simple as possible is generally a good idea.

When a linear relationship is assumed, the regression analysis is called *linear regression*. The simplest application of linear regression is modeling the relationship between two attributes: an input attribute X and a target attribute Y. In this simple linear-regression problem, the regression function has the following form:

$$Y = \omega_0 + \omega_1 X$$

This regression function is just the equation of a line (often written as $y = mx + c$) that is familiar to most people from high school geometry.[3] The variables ω_0 and ω_1 are the parameters of the regression function. Modifying these parameters changes how the function maps from the input X to the output Y. The parameter ω_0 is the y-intercept (or c in high school geometry) that specifies where the line crosses the vertical y axis when X is equal to zero. The parameter ω_1 defines the slope of the line (i.e., it is equivalent to m in the high school version).

In regression analysis, the parameters of a regression function are initially unknown. Setting the parameters of a regression function is equivalent to searching for the line that best fits the data. The strategy for setting these parameters begins by guessing parameters values and then iteratively updating the parameters so as to reduce the overall error of the function on the data set. The overall error is calculated in three steps:

1. The function is applied to the data set, and for each instance in the data set it estimates the value of the target attribute.

2. The error of the function for each instance is calculated by subtracting the estimated value of the target attribute from the actual value of the target attribute.

3. The error of the function for each instance is squared, and then these squared values are summed.

The error of the function for each instance is squared in step 3 so that the error in the instances where the function overestimates the target doesn't cancel out with the error when it underestimates. Squaring the error makes the error positive in both cases. This measure of error is known as the *sum of squared errors* (SSE), and the strategy of fitting a linear function by searching for the parameters that minimize the SSE is known as *least squares*. The SSE is defined as

$$SSE = \sum_{i=i}^{n} (target_i - prediction_i)^2$$

where the data set contains n instances, $target_i$ is the value of the target attribute for instance i in the data set, and $prediction_i$ is the estimate of the target by function for the same instance.

To create a linear-regression prediction model that estimates the likelihood of an individual's developing diabetes with respect to his BMI, we replace X with the BMI attribute, Y with the diabetes attribute, and apply the least-squares algorithm to find the best-fit line for the diabetes data set. Figure 11a illustrates this best-fit line and where it lies relative to the instances in the data set. In figure 11b, the dashed lines show the error (or residual) for each instance for this line. Using the least-squares approach, the best-fit line is the line that minimizes the sum of the squared residuals. The equation for this line is

$Diabetes = -7.38431 + 0.55593 * BMI$.

The slope parameter value $\omega_1 = 0.55593$ indicates that for each increase of one unit in BMI, the model increases the estimated likelihood of a person developing diabetes by a little more than half a percent. In order to predict the likelihood of a person's developing diabetes, we simply input his BMI into the model. For example, when BMI equals 20, the model returns a prediction of a 3.73 percent

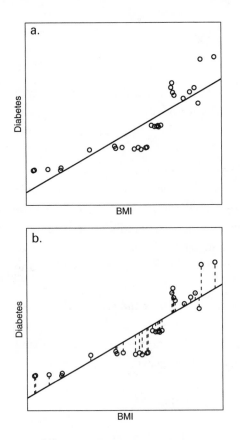

Figure 11 (*a*) The best-fit regression line for the model "Diabetes = −7.38431 + 0.55593 BMI." (*b*) The dashed vertical lines illustrate the residual for each instance.

likelihood for the diabetes attribute, and when BMI equals 21, the model predicts a 4.29 percent likelihood.[4]

Under the hood, a linear-regression model fitted using the least-squares method is actually calculating a weighted average over the instances. In fact, the intercept parameter value $\omega_0 = -7.38431$ ensures that the best-fit line goes through the point defined by the average BMI value and average diabetes value for the data set. If the average BMI value in the data set (BMI = 24.0932) is entered, the model estimates a 4.29 percent likelihood for the diabetes attribute, which is the average value for diabetes in the data set.

The weighting of the instances is based on the distance of the instance from the line: the farther an instance is away from the line, the larger the residual for that instance, and the algorithm will weight that instance by the residual squared. One consequence of this weighting is that instances that have extreme values (outliers) can have a disproportionately large impact on the line-fitting process, resulting in the line being dragged away from the other instances. Thus, it is important to check for outliers in a data set prior to fitting a line to the data set (or, in other words, training a linear regression function on the data set) using the least squares algorithm.

Linear-regression models can be extended to take multiple inputs. A new parameter is added to the model for each new input attribute, and the equation for the model is updated to include the result of multiplying the new attribute by the new parameter within the summation. For example, to extend the model to include the exercise and weight attributes as input, the structure of the regression function becomes

$$Diabetes = \omega_0 + \omega_1 BMI + \omega_2 Exercise + \omega_3 Weight.$$

In statistics, a regression function that maps from multiple inputs to a single output in this way is known as a *multiple linear regression function*. The structure of a multi-input regression function is the basis for a range of ML algorithms, including neural networks.

Correlation and regression are similar concepts insofar as both are techniques that focus on the relationship across columns in the data set. Correlation is focused on exploring whether a relationship exists between two attributes, and regression is focused on modeling an assumed relationship between attributes with the purpose of being able to estimate the value of one target attribute given the values of one or more input attributes. In the specific cases of Pearson correlation and linear regression, a Pearson correlation measures the degree to which two attributes have a linear relationship, and linear regression trained using least squares is a process to find the best-fit line that predicts the value of one attribute given the value of another.

Neural Networks and Deep Learning

A *neural network* consists of a set of neurons that are connected together. A neuron takes a set of numeric values as input and maps them to a single output value. At its core, a neuron is simply a multi-input linear-regression function. The only significant difference between the two is that in a neuron the output of the multi-input linear-regression function is passed through another function that is called the *activation function*.

These activation functions apply a nonlinear mapping to the output of the multi-input linear-regression function. Two commonly used activation functions are the *logistic function* and *tanh function* (see figure 12). Both functions take a single value x as input; in a neuron, this x value is the output from the multi-input linear-regression function the neuron has applied to its inputs. Also, both functions use Euler's number e, which is approximately equal to 2.71828182. These functions are sometimes called *squashing functions* because they take any value between plus infinity and minus infinity and map it into a small, predefined range. The output range of the logistic function is 0 to 1, and the tanh function is −1 to 1. As a consequence, the outputs of a neuron that uses a logistic function as its activation function are always between 0 and 1. The fact that both the logistic and tanh functions apply nonlinear mappings is clear in the S shape of the curves. The reason for introducing a nonlinear mapping

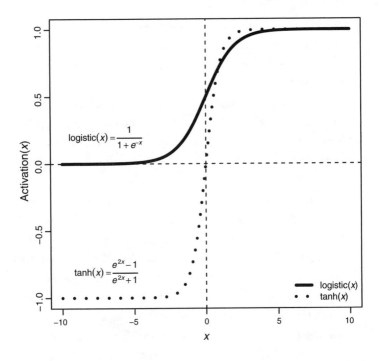

Figure 12 Mapping the logistic and tanh functions as applied to the input x.

into a neuron is that one of the limitations of a multi-input linear-regression function is that the function is by definition linear, and if all the neurons within a network implement only linear mappings, then the overall network is also limited to learning a linear functions. However, introducing a nonlinear activation function in the neurons of a network allows the network to learn more complex (nonlinear) functions.

It is worth emphasizing that each neuron in a neural network is doing a very simple set of operations:

1. Multiplying each input by a weight.

2. Adding together the results of the multiplications.

3. Pushing this result through an activation function.

Operations 1 and 2 are just the calculation of a multi-input regression function over the inputs, and operation 3 is the application of the activation function.

All the connections between the neurons in a neural network are directed and have a weight associated with them. The weight on a connection coming into a neuron is the weight that the neuron applies to the input it receives on that connection when it is calculating the multi-input regression function over its inputs. Figure 13 illustrates the topological structure of a simple neural network. The squares on the left side of the figure, labeled A and B, represent locations in memory that we use to present input data to the network. No processing or transformation of data is carried out at these locations. You can think of these nodes as input or sensing neurons, whose output activation is set to the value of the input.[5] The circles in figure 13 (labeled C, D, E, and F) represent the neurons

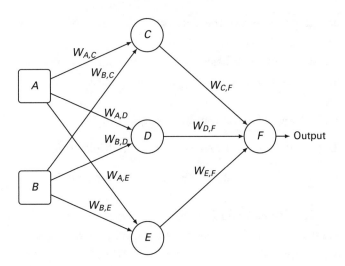

Figure 13 A simple neural network.

in the network. It is often useful to think of the neurons in a network as organized into layers. This network has three layers of neurons: the input layer contains A and B; one hidden layer contains C, D, and E; and the output layer contains F. The term *hidden layer* describes the fact that the neurons in a layer are in neither the input layer nor the output layer, so in this sense they are hidden from view.

The arrows connecting the neurons in the network represent the flow of information through the network. Technically, this particular network is a feed-forward neural network because there are no loops in the network: all the connections point forward from the input toward the output. Also, this network is considered fully connected because each neuron is connected to all the neurons in the next layer in the network. It is possible to create many different types of neural networks by changing the number of layers, the number of neurons in each layer, the type of activation functions used, the direction of the connections between layers, and other parameters. In fact, much of the work involved in developing a neural network for a particular task involves experimenting to find the best network layout for that task.

The labels on each arrow represent the weight that the node at the end of the arrow applies to the information passed along that connection. For example, the arrow connecting C with F indicates that the output from C is passed as an input to F, and F will apply the weight $W_{C,F}$ to the input from C.

If we assume that the neurons in the network in figure 13 use a tanh activation function, then we can define the calculation carried out in neuron F of the network as

$$Output = \tanh\left(\omega_{C,F}C + \omega_{D,F}D + \omega_{E,F}E\right)$$

The mathematical definition of the processing carried out in neuron F shows that the final output of the network is calculated using a composition of a set of functions. The phrase "composing functions" just means that the output of one function is used as input to another function. In this case, the outputs of neurons C, D, and E are used as inputs to neuron F, so the function implemented by F composes the functions implemented by C, D, and E.

Figure 14 makes this description of neural networks more concrete, illustrating a neural network that takes a person's body-fat percentage and VO_2 max (a measure of the maximum amount of oxygen that a person can use in a minute) as input and calculates a fitness level for the that person.[6] Each neuron in the middle layer of the network calculates a function based on the body-fat percentage and VO_2 max: $f_1()$, $f_2()$, and $f_3()$. Each of these functions models the interaction between the inputs in a different way. These functions essentially represent new attributes that are derived from the raw inputs to the network. They are similar to the BMI attribute described earlier, which was calculated as a function of weight and height. Sometimes it is possible to interpret what the output of a neuron in the network represents insofar as it is possible to provide a domain-theoretic description of what the derived attribute represents and to understand why this derived attribute is useful to the network. Often, however, the derived attribute calculated by a neuron will not have a symbolic meaning for humans. These attributes are instead capturing interactions between the other attributes that the network has found to be useful. The final node in the network, f_4, calculates another function—over the outputs of $f_1()$,

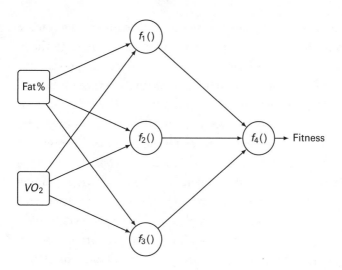

Figure 14 A neural network that predicts a person's fitness level.

$f_2()$, and $f_3()$—the output of which is the fitness prediction returned by the network. Again, this function may not be meaningful to humans beyond the fact that it defines an interaction the network has found to have a high correlation with the target attribute.

Training a neural network involves finding the correct weights for the connections in the network. To understand how to train a network, it is useful to begin by thinking about how to train the weights for a single neuron in the output layer of the network. Assume that we have a training data set that has both inputs and target output for each instance. Also, assume that the connections coming into the neuron already have weights assigned to them. If we take an instance from the data set and present the values of the input attributes for this instance to the network, the neuron will output a prediction for the target. By subtracting this prediction from the value for the target in the data set, we can measure the neuron's error on that instance. Using some basic calculus, it is possible to derive a rule to update the weights on the connections coming into a neuron given a measure of the neuron's output error so as to reduce the neuron's error. The precise definition of this rule will vary depending on the activation function used by the neuron because the activation function affects the derivative used in the derivation of the rule. But we can give the following intuitive explanation of how the weight-update rule works:

1. If the error is 0, then we should not change the weights on the inputs.

2. If the error is positive, we will decrease the error if we increase the neuron's output, so we must increase the weights for all the connections where the input is positive and decrease the weights for the connections where the input is negative.

3. If the error is negative, we will decrease the error if we decrease the neuron's output, so we must decrease the

weights for all the connections where the input is positive and increase the weights for the connections where the input is negative.

The difficulty in training a neural network is that the weight-update rule requires an estimate of the error at a neuron, and although it is straightforward to calculate the error for each neuron in the output layer of the network, it is difficult to calculate the error for the neurons in the earlier layers. The standard way to train a neural network is to use an algorithm called the *backpropagation algorithm* to calculate the error for each neuron in the network and then use the weight-update rule to modify the weights in the network.[7] The backpropagation algorithm is a supervised ML algorithm, so it assumes a training data set that has both inputs and the target output for each instance. The training starts by assigning random weights to each of the connections in the network. The algorithm then iteratively updates the weights in the network by showing training instances from the data set to the network and updating the network weights until the network is working as expected. The algorithm's name comes from the fact that after each training instance is presented to the network, the algorithm passes (or backpropagates) the error of the network back through the network starting at the output layer and at each layer in the network calculates the error for the neurons in that layer before sharing this error back to the neurons in the preceding layer. The main steps in the algorithm are as follows:

1. Calculate the error for the neurons in the output layer and use the weight-update rule to update the weights coming into these neurons.

2. Share the error calculated at a neuron with each of the neurons in the preceding layer that is connected to that neuron in proportion to the weight of the connection between the two neurons.

3. For each neuron in the preceding layer, calculate the overall error of the network that the neuron is responsible for by summing the errors that have been backpropagated to it and use the result of this error summation to update the weights on the connections coming into this neuron.

4. Work back through the rest of the layers in the network by repeating steps 2 and 3 until the weights between the input neurons and the first layer of hidden neurons have been updated.

In backpropagation, the weight updates for each neurons are scaled to reduce but not to eliminate the neuron's error in the training instance. The reason for this is that the goal of training the network is to enable it to generalize to new instances that are not in the training data rather than to memorize the training data. So each set of weight updates nudges the network toward a set of weights that are generally better over the whole data set, and over many iterations the network converges on a set of weights that captures the general distribution of the data rather than the specifics of the training instances. In some versions of backpropagation, the weights are updated after a number of instances (or batch of instances) have been presented to the network rather than after each training instance. The only adjustment required in these versions is that the algorithm uses the average error of the network on a batch as the measure of error at the output layer for the weight-update process.

One of the most exciting technical developments in the past 10 years has been the emergence of deep learning. *Deep-learning networks* are simply neural networks that have multiple[8] layers of hidden units; in other words, they are *deep* in terms of the number of hidden layers they have. The neural network in figure 15 has five layers: one input layer on the left containing three neurons, three hidden layers (the black circles), and one output layer on the right containing two neurons. This network illustrates that it is possible to have a different number of neurons in each layer: the input layer has three neurons; the first hidden layer has five; each of the next two hidden layers has four; and the output layer has two. This network also shows that it is possible to have multiple

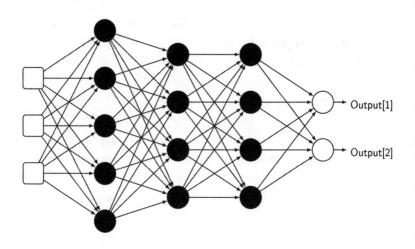

Figure 15 A deep neural network.

neurons in the output layer. Using multiple output neurons is useful if the target is a nominal or ordinal data type that has distinct levels. In these scenarios, the network is set up so that there is one output neuron for each level, and the network is trained so that for each input only one of the output neurons outputs a high activation (denoting the predicted target level).

As in the previous networks we have looked at, the one shown in figure 15 is a fully connected, feed-forward network. However, not all networks are fully connected, feed-forward networks. In fact, myriad network topologies have been developed. For example, *recurrent neural networks* (RNNs) introduce loops in the network topology: the output of a neuron for one input is fed back into the neuron during the processing of the next input. This loop gives the network a memory that enables it to process each input in the context of the previous inputs it has processed. As a consequence, RNNs are suitable for processing sequential data such as language.[9] Another well-known deep neural network architecture is a *convolutional neural network* (CNN). CNNs were originally designed for use with image data (Le Cun 1989). A desirable characteristic of an image-recognition network is that it should be able to recognize if a visual feature has occurred in an image irrespective of where in the image it has occurred. For example, if a network is doing face recognition, it needs to be able to recognize the shape of an eye whether the eye is in the top-right corner of the image or in the center of the image. CNNs achieve this by having groups of neurons that share the same set of weights on their inputs. In this context, think of a set of input weights as defining a function that returns true if a particular visual feature occurs in the set of pixels that are passed into the function. This means that each group of neurons that share their weights learns to identify a particular visual feature, and each neuron in the group acts as a detector for that feature. In a CNN, the neurons within each group are arranged so that each neuron examines a different location in the image, and the group covers the entire image. As a consequence, if the visual feature the group detects occurs anywhere in the image, one of the neurons in the group will identify it.

The power of deep neural networks comes from the fact that they can automatically learn useful attributes, such as the feature detectors in CNNs. In fact, deep learning is sometimes known as *representation learning* because these deep networks are essentially learning a new representation of the input data that is better at predicting the target output attribute than the original raw input is. Each neuron in a network defines a function that maps the values coming into the neuron into a new output attribute. So a neuron in the first layer of a network might learn a function that maps the raw input values (such as weight and height) into an attribute that is more useful than individual input values (such as BMI). However, the output of this neuron, along with the outputs of its sister neurons in the first layer, is then fed into the neurons in the second layer, and these second-layer neurons try to learn functions that map the outputs of the first layer into new and yet more useful representations. This process of mapping inputs to new attributes and feeding these new attributes as inputs to new functions continues throughout the network, and as a network gets deeper, it can learn more and more complex mappings from raw inputs to new attribute representations. It is the ability to automatically learn complex mappings of input data to useful attribute representations that has made deep-learning models so accurate in tasks with high-dimensional inputs (such as image and text processing).

It has been known for a long time that making neural networks deeper allows the network to learn more complex mappings of data. The reason that deep learning has not really taken off until the past few years, however, is that the standard combination of using a random-weight initialization followed by the backpropagation algorithm doesn't work well with deep networks. One problem with the backpropagation algorithm is that the error gets shared out as the process goes back through the layers, so that in a deep network by the time the algorithm reaches the early layers of the network, the error estimates are not that useful anymore.[10] As a result, the layers in the early parts of the network don't learn useful transformations for the data. In the past few years, however, researchers have developed new types of neurons and adaptations to the backpropagation algorithm that deal with this problem. It has also been found that being careful with how the network weights are initialized is also helpful. Two other factors that formerly made training deep networks difficult were that it takes a great deal of computing power to train a neural network, and neural networks work best when there is a great deal of training data. However, as we have already discussed, in recent years significant increases in the availability of computing power and large data sets have made the training of deep networks more feasible.

Decision Trees

Linear regression and neural networks work best with numeric inputs. If the input attributes in a data set are primarily nominal or ordinal, however, then other ML algorithms and models, such as *decision trees*, may be more appropriate.

A decision tree encodes a set of *if then, else* rules in a tree structure. Figure 16 illustrates a decision tree for deciding whether an email is spam or not. Rectangles with rounded corners represent tests on attributes, and the square nodes indicate decision, or classification, nodes. This tree encodes the following rules: *if the email is from*

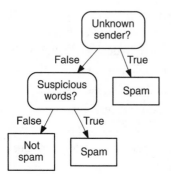

Figure 16   A decision tree for determining whether an email is spam or not.

an unknown sender, then it is spam; if it isn't from an unknown sender but contains suspicious words, then it is spam; if it is neither from an unknown sender nor contains suspicious words, then it is not spam. In a decision tree, the decision for an instance is made by starting at the top of the tree and navigating down through the tree by applying a sequence of attribute tests to the instance. Each node in the tree specifies one attribute to test, and the process descends the tree node by node by choosing the branch from the current node with the label matching the value of the test attribute of the instance. The final decision is the label of the terminating (or *leaf*) node that the instance descends to.

Each path in a decision tree, from root to leaf, defines a classification rule composed of a sequence of tests. The goal of a decision-tree-learning algorithm is to find a set of classification rules that divide the training data set into sets of instances that have the same value for the target attribute. The idea is that if a classification rule can separate out from a data set a subset of instances that have the same target value, and if this classification rule is true for a new example (i.e., the example goes down that path in the tree), then it is likely that the correct prediction for this new example is the target value shared by all the training instances that fit this rule.

The progenitor of most modern ML algorithms for decision-tree learning is the *ID3 algorithm* (Quinlan 1986).

ID3 builds a decision tree in a recursive, depth-first manner, adding one node at a time, starting with the root node. It begins by selecting an attribute to test at the root node. A branch is grown from the root for each value in the domain of this test attribute and is labeled with that value. For example, a node with a binary test attribute will have two branches descending from it. The data set is then divided up: each instance in the data set is pushed down the branch and given a label that matches the value of the test attribute for the instance. ID3 then grows each branch using the same process used to create the root node: select a test attribute, add a node with branches, split the data by funneling the instances down the relevant branches. This process continues until all the instances on a branch have the same value for the target attribute, in which case a terminating node is added to the tree and labeled with the target attribute value shared by all the instances on the branch.[11]

ID3 chooses the attribute to test at each node in the tree so as to minimize the number of tests required to create pure sets (i.e., sets of instances that have the same value for the target attribute). One way to measure the purity of a set is to use Claude Shannon's *entropy* metric. The minimum possible entropy for a set is zero, and a pure set has an entropy of zero. The numeric value of the maximum possible entropy for a set depends on the size of the set and the number of different types of elements that can be in the set. A set will have maximum entropy when all the elements in it are of different types.[12] ID3 selects the attribute to test at a node to be the attribute that results in the lowest-weighted entropy after splitting the data set at the node using this attribute. The weighted entropy for an attribute is calculated by (1) splitting the data set using the attribute; (2) calculating the entropy of the resulting sets; (3) weighting each of these entropies by the fraction of data that is in the set; and (4) then summing the results.

Table 3 lists a data set of emails in which each email is described by a number of attributes and whether it is a spam email or not. The "attachment" attribute is true for emails that have an attachment and false otherwise (in this sample of emails, none of the emails has an attachment).

Table 3 A Data Set of Emails: Spam or Not Spam?

Attachment	Suspicious Words	Unknown Sender	Spam
False	False	True	True
False	False	True	True
False	True	False	True
False	False	False	False
False	False	False	False

The "suspicious words" attribute is true if the email contains one or more words on a predefined list of suspicious words. The "unknown sender" attribute is true if the sender of the email is not in the recipient's address book. This is the data set that was used to train the decision tree shown in figure 16. In this data set, the attributes "attachment," "suspicious words," and "unknown sender" are the input attributes and the "spam" attribute is the target. The "unknown sender" attribute splits the data set into purer sets more than any of the other attributes does (one set containing instances where "Spam = True" and another set in which "Spam = False" for the majority of instances). As a consequence, "unknown sender" is put at the root node (see figure 17). After this initial split, all of the instances

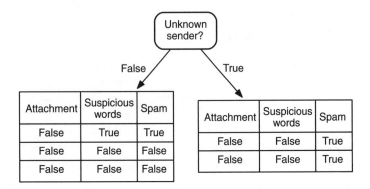

Figure 17 Creating the root node in the tree.

on the right branch have the same target value. However, the instances on the left branch have different values for the target. Splitting the instances on the left branch using the "suspicious words" attribute results in two pure sets: one where "Spam = False" and another where "Spam = True." So "suspicious words" is selected as the test attribute for a new node on the left branch (see figure 18). At this point, the data subsets at the end of each branch are pure, so the algorithm finishes and returns the decision tree shown in figure 16.

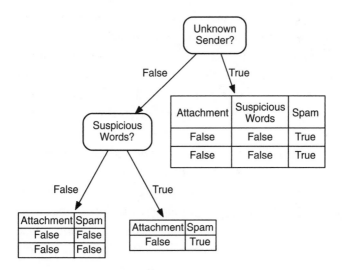

Figure 18 Adding the second node to the tree.

One of the strengths of decision trees is that they are simple to understand. Also it is possible to create very accurate models based on decision trees. For example, a *random-forest model* is composed of a set of decision trees, where each tree is trained on a random subsample of the training data, and the prediction returned by the model for an individual query is the majority prediction across all the trees in the forest. Although decision trees work well with both nominal and ordinal data, they struggle with numeric data. In a decision tree, a separate branch descends from each node for each value in the domain of the attribute tested at the node. Numeric attributes, however, have an infinite number of values in their domains, with the implication that a tree would need an infinite number of branches. One solution to this problem is to transform numeric attributes into ordinal attributes, although doing so requires the definition of appropriate thresholds, which can also be difficult.

Finally, because a decision-tree-learning algorithm repeatedly divides a data set as a tree becomes large, it becomes more sensitive to noise (such as mislabeled instances). The subset of examples on each branch becomes smaller, and so the data sample each classification rule is based on becomes smaller. The smaller the data sample used to define a classification rule, the more sensitive to noise the rule becomes. As a consequence, it is a good idea to keep decision trees shallow. One approach is to stop the growth of a branch when the number of instances on the branch is still less than a predefined threshold (e.g., 20 instances). Other approaches allow the tree to grow and then prune the tree back. These approaches typically use statistical tests or the performance of the model on a set of instances specifically chosen for this task to identify splits near the bottom of the tree that should be removed.

Bias in Data Science

The goal of ML is to create models that encode appropriate generalizations from data sets. Two major factors contribute to the generalization (or model) that an ML algorithm will generate from a data set. The first is the data set the algorithm is run on. If the data set is not representative of the population, then the model the algorithm generates won't be accurate. For example, earlier we developed a regression model that predicted the likelihood that an individual will develop Type 2 diabetes based on his BMI. This model was generated from a data set of American white males. As a consequence, this model is unlikely to be accurate if used to predict the likelihood of diabetes for females or for males of different race or ethnic backgrounds. The term *sample bias* describes how the process used to select a data set can introduce biases into later analysis, be it a statistical analysis or the generation of predictive models using ML.

The second factor that affects the model generated from a data set is the choice of ML algorithm. There are many different ML algorithms, and each one encodes a different way to generalize from a data set. The type of generalization an algorithm encodes is known as the *learning bias* (or sometimes the *modeling* or *selection bias*) of the algorithm. For example, a linear-regression algorithm encodes a linear generalization from the data and as a result ignores nonlinear relationships that may fit the data more closely. Bias is normally understood as a bad thing. For example, the sampling bias is a bias that a data scientist will try to avoid. However, without a learning bias there can be no learning, and the algorithm will only be able to memorize the data.

However, because ML algorithms are biased to look for different types of patterns, and because there is no one learning bias across all situations, there is no one best ML algorithm. In fact, a theorem known as the *"no free lunch theorem"* (Wolpert and Macready 1997) states that there is no one best ML algorithm that on average outperforms all other algorithms across all possible data sets. So the modeling phase of the CRISP-DM process normally involves building multiple models using different algorithms and comparing the models to identify which algorithm generates the best model. In effect, these experiments are testing which learning bias on average produces the best models for the given data set and task.

Evaluating Models: Generalization Not Memorization

Once a data scientist has selected a set of ML algorithms to experiment with on a data set, the next major task is to create a test plan for how the models generated by these algorithms will be evaluated. The goal of the test plan is to ensure that the evaluation provides realistic estimates of model performance on unseen data. A prediction model that simply memorizes a data set is unlikely to do a good job at estimating values for new examples. One problem with just memorizing data is that most data sets will contain noise. So a prediction model that merely memorizes data is also memorizing the noise in the data. Another problem with just memorizing the data is that it reduces the prediction process to a table lookup and leaves unsolved the problem of how to generalize from the training data to new examples that aren't in the table.

One part of the test plan relates to how the data set is used to train and test the models. The data set has to be used for two different purposes. The first is to find which algorithm generates the best models. The second is to estimate the generalization performance of the best model—that is, how well the model is likely to do on unseen data. The golden rule for evaluating models is that models should never be tested on the same data they were trained on. Using the same data for training and testing models is equivalent to giving a class of students the

The golden rule for evaluating models is that models should never be tested on the same data they were trained on.

questions on an exam the night before the test is held. The students will of course do very well in the test, and their scores will not reflect their real proficiency with the general course material. So, too, with ML models: if a model is evaluated on the same data that it is trained on, the results of the evaluation will be optimistic compared to the model's real performance. The standard process for ensuring that the models aren't able to peek at the test data during training is to split the data into three parts: a training set, a validation set, and a test set. The proportions of the split will vary between projects, but splits of 50:20:30 and 40:20:40 are common. The size of the data set is a key factor in determining the splits: generally, the larger the data set, the larger the test set. The training set is used to train an initial set of models. The validation set is then used to compare the performance of these models on unseen data. Comparing the performance of these initial models on the validation set enables us to determine which algorithm generated the best model. Once the best algorithm has been selected, the training and validation set can be merged back together into a larger training set, and this data set is fed into the best algorithm to create the final model. It is crucial that the test set is not used during the process to select the best algorithm, nor should it be used to train this final model. If these caveats are followed, then the test set can be used to estimate the generalization performance of this final model on unseen data.

The other major component of the test plan is to choose the appropriate evaluation metrics to use during the testing. In general, models are evaluated based on how often the outputs of the model match the outputs listed in the test set. If the target attribute is a numeric value, then the sum of squared errors is one way to measure the accuracy of a model on the test set. If the target attribute is nominal or ordinal, then the simplest way to estimate the model accuracy is to calculate the proportion of examples of the test set the model got correct. However, in some contexts it is important to include an error analysis within the evaluation. For example, if a model is used in a medical diagnosis setting, it is much more serious if the model diagnoses an ill patient as healthy than if it diagnoses a healthy patient as ill. Diagnosing an ill patient as healthy may result in the patient being sent home without receiving appropriate medical attention, but if a model diagnoses a healthy patient as ill, this error is likely to be discovered through later testing the patient will receive. So the evaluation metric used to evaluate these types of models should weight one type of error more than the other when estimating model performance. Once the test plan has been created, the data scientist can begin training and evaluating models.

Summary

This chapter started by noting that data science is a partnership between a data scientist and a computer. Machine learning provides a set of algorithms that generate models from a large data set. However, whether these models are useful will depend on the data scientist's expertise. For a data science project to succeed, the data set should be representative of the domain and should include relevant attributes. The data scientist should evaluate a range of ML algorithms to find which one generates the best models. The model-evaluation process should follow the golden rule that a model should never be evaluated on the data it was trained on.

Currently in most data science projects, the primary criterion for selecting which model to use is model accuracy. However, in the near future, data usage and privacy regulations may affect the selection of ML algorithms. For example, the General Data Protection Regulations will come into force in the European Union on May 25, 2018. We discuss these regulations in relation to data usage in chapter 6, but for now we just want to point out that some articles in the regulations may appear to mandate a "right to explanation" in relation to automated decision processes.[13] A potential implication of such a right is that using models, such a neural networks, that are difficult to interpret for decisions relating to individuals may become problematic. In such circumstances, the transparency and ease of explanation of some models, such as decision trees, may make the use of these models more appropriate.

Finally, the world changes, and models don't. Implicit in the ML process of data set construction, model training, and model evaluation is the assumption that the future will be the same as the past. This assumption is known as the *stationarity assumption*: the processes or behaviors that are being modeled are stationary through time (i.e., they don't change). Data sets are intrinsically historic in the sense that data are representations of observations that were made in the past. So, in effect, ML algorithms search through the past for patterns that might generalize to the future. Obviously, this assumption doesn't always hold. Data scientists use the term *concept drift* to describe how a process or behavior can change, or drift, as time passes. This is why models go out of date and need to be retrained and why the CRISP-DM process includes the outer circle shown in figure 4 to emphasize that data science is iterative. Processes need to put in place postmodel deployment to ensure that a model has not gone stale, and when it has, it should be retrained. The majority of these decisions cannot be automated and require human insight and knowledge. A computer will answer the question it is posed, but unless care is taken, it is very easy to pose the wrong question.

STANDARD DATA SCIENCE TASKS

One of the most important skills for a data scientist is the ability to frame a real-world problem as a standard data science task. Most data science projects can be classified as belonging to one of four general classes of task:

- Clustering (or segmentation)

- Anomaly (or outlier) detection

- Association-rule mining

- Prediction (including the subproblems of classification and regression)

Understanding which task a project is targeting can help with many project decisions. For example, training a prediction model requires that each of the instances in the data set include the value of the target attribute. So knowing that the project is doing prediction gives guidance (through requirements) in terms of data set design. Understanding the task also informs which ML algorithm(s) to use. Although there are a large number of ML algorithms, each algorithm is designed for a particular data-mining task. For example, ML algorithms that generate decision-tree models are designed primarily for prediction tasks. There is a many-to-one relationship between ML algorithms and a task, so knowing the task doesn't tell you exactly which algorithm to use, but it does define a set of algorithms that are designed for the task. Because the data science task affects both the data set design and the selection of ML algorithms, the decision regarding which task the project will target has to be made early on in the project life cycle, ideally during the business-understanding phase of the CRISP-DM life cycle. To provide a better understanding of each of these tasks, this chapter describes how some standard business problems map to tasks.

Who Are Our Customers? (Clustering)

One of the most frequent application areas of data science in business is to support marketing and sales campaigns. Designing a targeted marketing campaign requires an understanding of the target customer. Most businesses have a diverse range of customers with a variety of needs, so using a one-size-fits-all approach is likely to fail with a large segment of a customer base. A better approach is to try to identify a number of customer personas or customer profiles, each of which relates to a significant segment of the customer base, and then to design targeted marketing campaigns for each persona. These personas can be created using domain expertise, but it is generally a good idea to base the personas on the data that the business has about its customers. Human intuition about customers can often miss important nonobvious segments or not provide the level of granularity that is required for nuanced marketing. For example, Meta S. Brown (2014) reports how in one data science project the well-known stereotype *soccer mom* (a suburban homemaker who spends a great deal of time driving her children to soccer or other sports practice) didn't resonate with a customer base. However, using a data-driven clustering process identified more focused personas, such as *mothers working full-time outside the home with young children in daycare* and *mothers who work part-time with high-school-age children* and *women interested in food and health and who do not have children*. These customer personas define clearer targets for marketing campaigns and may highlight previously unknown segments in the customer base.

The standard data science approach to this type of analysis is to frame the problem as a *clustering* task. Clustering involves sorting the instances in a data set into

Human intuition about customers can often miss important nonobvious segments or not provide the level of granularity that is required for nuanced marketing.

subgroups containing similar instances. Usually clustering requires an analyst to first decide on the number of subgroups she would like identified in the data. This decision may be based on domain knowledge or informed by project goals. A clustering algorithm is then run on the data with the desired number of subgroups input as one of the algorithms parameters. The algorithm then creates that number of subgroups by grouping instances based on the similarity of their attribute values. Once the algorithm has created the clusters, a human domain expert reviews the clusters to interpret whether they are meaningful. In the context of designing a marketing campaign, this review involves checking whether the groups reflect sensible customer personas or identifies new personas not previously considered.

The range of attributes that can be used to describe customers for clustering is vast, but some typical examples include demographic information (age, gender, etc.), location (ZIP code, rural or urban address, etc.), transactional information (e.g., what products or services they have purchased), the revenue the company generates from them, how long have they been customers, if they are a member of a loyalty-card scheme, whether they ever returned a product or made a complaint about a service, and so on. As is true of all data science projects, one of the biggest challenges with clustering is to decide which attributes to include and which to exclude so as to get the best results.

Making this decision on attribute selection will involve iterations of experiments and human analysis of the results of each iteration.

The best-known ML algorithm for clustering is the *k-means* algorithm. The k in the name signals that the algorithm looks for k clusters in the data. The value of k is predefined and is often set through a process of trial-and-error experimentation with different values of k. The k-means algorithm assumes that all the attributes describing the customers in the data set are numeric. If the data set contains nonnumeric attributes, then these attributes need to be mapped to numeric values in order to use k-means, or the algorithm will need to be amended to handle these nonnumeric values. The algorithm treats each customer as a point in a point cloud (or scatterplot), where the customer's position is determined by the attribute values in her profile. The goal of the algorithm is to find the position of each cluster's center in the point cloud. There are k clusters, so there are k cluster centers (or means)— hence the name of the algorithm.

The k-means algorithm begins by selecting k instances to act as initial cluster centers. Current best practice is to use an algorithm called "k-means++" to select the initial cluster centers. The rationale behind k-means++ is that it is a good idea to spread out the initial cluster centers as much as possible. So in k-means++ the first cluster center is set by randomly selecting one of the instances in

As is true of all data science projects, one of the biggest challenges with clustering is to decide which attributes to include and which to exclude so as to get the best results.

the data set. The second and subsequent cluster centers are set by selecting an instance from the data set with the probability that an instance selected is proportional to the squared distance from the closest existing cluster center. Once all k cluster centers have been initialized, the algorithm works by iterating through a two-step process: first, assigning each instance to the nearest cluster center, and then, second, updating the cluster center to be in the middle of the instances assigned to it. In the first iteration the instances are assigned to the nearest cluster center returned by the k-means++ algorithm, and then these cluster centers are moved so that they are positioned at the center of instances assigned to them. Moving the cluster centers is likely to move them closer to some instances and farther away from other instances (including farther away from some instances assigned to the cluster center). The instances are then reassigned, again to the closest updated cluster center. Some instances will remain assigned to the same cluster center, and others may be reassigned to a new cluster center. This process of instance assignment and center updating continues until no instances are assigned to a new cluster center during an iteration. The k-means algorithm is nondeterministic, meaning that different starting positions for the cluster centers will likely produce different clusters. As a result, the algorithm is typically run several times, and the results of these different runs are then compared to see which clusters appear most sensible given the data scientist's domain knowledge and understanding.

When a set of clusters for customer personas has been deemed to be useful, the clusters are often given names to reflect the main characteristics of the cluster persona. Each cluster center defines a different customer persona, with the persona description generated from the attribute values of the associated cluster center. The k-means algorithm is not required to return equal-size clusters, and, in fact, it is likely to return different-size clusters. The sizes of the clusters can be useful, though, because they can help to guide marketing. For example, the clustering process may reveal small, focused clusters of customers that current marketing campaigns are missing. Or an alternative strategy might be to focus on clusters that contain customers that generate a great deal of revenue. Whatever marketing strategy is adopted, understanding the segments within a customer base is the prerequisite to marketing success.

One of the advantages of clustering as an analytics approach is that it can be applied to most types of data. Because of its versatility, clustering is often used as a data-exploration tool during the data-understanding stage of many data science projects. Also, clustering is also useful across a wide range of domains. For example, it has been used to analyze students in a given course in order to identify groups of students who need extra support or prefer different learning approaches. It has also been used to identify groups of similar documents in a corpus, and in science it has been used in bio-informatics to analyze gene sequences in microarray analysis.

Is This Fraud? (Anomaly Detection)

Anomaly detection or outlier analysis involves searching for and identifying instances that do not conform to the typical data in a data set. These nonconforming cases are often referred to as *anomalies* or *outliers*. Anomaly detection is often used in analyzing financial transactions in order to identify potential fraudulent activities and to trigger investigations. For example, anomaly detection might uncover fraudulent credit card transactions by identifying transactions that have occurred in an unusual location or that involve an unusually large amount compared to other transactions on a particular credit card.

The first approach that most companies typically use for anomaly detection is to manually define a number of rules based on domain expertise that help with identifying anomalous events. This rule set is often defined in SQL or in another language and is run against the data in the business databases or data warehouse. Some programming languages have begun to include specific commands to facilitate the coding of these types of rules. For example, database implementations of SQL now includes a MATCH_RECOGNIZE function to facilitate pattern matching in data. A common pattern in credit card fraud is that when a credit card gets stolen, the thief first checks that the card is working by purchasing a small item on the card, and then if that transaction goes through, the thief as quickly as possible follows that purchase with the purchase of an expensive item before the card is canceled. The MATCH_RECOGNIZE function in SQL enables database programmers to write scripts that identify sequences of transactions on a credit card that fit this pattern and either block the card automatically or trigger a warning to the credit-card company. Over time, as more anomalous transactions are identified—for example, by customers reporting fraudulent transactions—the set of rules identifying anomalous transactions is expanded to handle these new instances.

The main drawback with a rule-based approach to anomaly detection is that defining rules in this way means that anomalous events can be identified only after they have occurred and have come to the company's attention. Ideally, most organizations would like to be able to identify anomalies when they first happen or if they have happened but have not been reported. In some ways, anomaly detection is the opposite of clustering: the goal of clustering is to identify groups of similar instances, whereas the goal of anomaly detection is to find instances that are dissimilar to the rest of the data in the data set. By this intuition, clustering can also be used to automatically identify anomalies. There are two approaches to using clustering for anomaly detection. The first is that the normal data will be clustered together, and the anomalous records will be in separate clusters. The clusters containing the anomalous records will be small and so will be clearly distinct from the large clusters for the main body of the records. The second approach is to measure the distance between each instance and the center of the cluster. The farther away the instance is from the center of the cluster, the more likely it is to be anomalous and thus to need investigation.

Another approach to anomaly detection is to train a prediction model, such as a decision tree, to classify instances as anomalous or not. However, training such a model normally requires a training data set that contains both anomalous records and normal records. Also, it is not enough to have just a few instances of anomalous records; in order to train a normal prediction model, the data set needs to contain a reasonable number of instances from each class. Ideally, the data set should be balanced; in a binary-outcome case, balance would imply a 50:50 split in the data. In general, acquiring this type of training data for anomaly detection is not feasible: by definition, anomalies are rare events, occurring maybe in 1 to 2 percent or less of the data. This data constraint precludes the use of normal, off-the-shelf prediction models. There are, however, ML algorithms known as *one-class classifiers* that are designed to deal with the type of imbalanced data that are typical of anomaly-detection data sets.

The *one-class support-vector machine* (SVM) algorithm is a well-known one-class classifier. In general terms, the one-class SVM algorithm examines the data as one unit (i.e., a single class) and identifies the core characteristics and expected behavior of the instances. The algorithm will then indicate how similar or dissimilar each instance is from the core characteristics and expected behavior. This information can then be used to identify instances that warrant further investigation (i.e., the anomalous records). The more dissimilar an instance is, the more likely that it should be investigated.

The fact that anomalies are rare means that they can be easy to miss and difficult to identify. As a result, data scientists often combine a number of different models to detect anomalies. The idea is that different models will capture different types of anomalies. In general, these models are used to supplement the known rules within the business that already define various types of anomalous activity. The different models are integrated together into a decision-management solution that enables the predictions from each of the models to feed into a decision of the final predicted outcome. For example, if a transaction is identified as fraudulent by only one out of four models, the decision system may decide that it isn't a true case of fraud, and the transaction can be ignored. Conversely, however, if three or four out of the four models have identified the transaction as possible fraud, then the transaction would be flagged for a data scientist to investigate.

Anomaly detection can be applied to many problem domains beyond credit card fraud. More generally, it is used by clearinghouses to identify financial transactions that require further investigation as potentially fraudulent or as cases of money laundering. It is used in insurance-claims analysis to identify claims that are not in keeping with a company's typical claims. In cybersecurity, it is used to identify network intrusions by detecting possible hacking or untypical behavior by employees. In the medical domain, identifying anomalies in medical records can be useful for diagnosing disease and in studying treatments and their effects on the body. Finally, with the proliferation of sensors and the increasing usage of Internet of Things technology, anomaly detection will play an important role in monitoring data and alerting us when abnormal sensor events occur and action is required.

Do You Want Fries with That? (Association-Rule Mining)

A standard strategy in sales is cross-selling, or suggesting to customers who are buying products that they may also want to purchase other related or complementary products. The idea is to increase the customers' overall spending by getting them to purchase more products and at the same time to improve customer service by reminding customers of products they probably wanted to buy but may have forgotten to do so. The classic example of the cross-sell is when a waiter in a hamburger restaurant asks a customer who has just ordered a hamburger, "Do you want fries with that?" Supermarkets and retailer businesses know that shoppers purchase products in groups, and they use this information to set up cross-selling opportunities. For example, supermarket customers who buy hot dogs are also likely to purchase ketchup and beer. Using this type of information, a store can plan the layout of the products. Locating hot dogs, ketchup, and beer near each other in the store helps customers to collect this group of items quickly and may also boost the store sales because customers who are purchasing hot dogs might see and purchase the ketchup and beer that they forgot they needed. Understanding these types of associations between products is the basis of all cross-selling.

Association-rule mining is an unsupervised-data-analysis technique that looks to find groups of items that frequently co-occur together. The classic case of association mining is *market-basket analysis*, wherein retail companies try to identify sets of items that are purchased together, such as hot dogs, ketchup, and beer. To do this type of data analysis, a business keeps track of the set (or basket) of items that each customer bought during each visit to the store. Each row in the data set describes one basket of goods purchased by a particular customer on a particular visit to the store. So the attributes in the data set are the products the store sells. Given these data, association-rule mining looks for items that co-occur within each basket of goods. Unlike clustering and anomaly detection, which focus on identifying similarities or differences between instances (or rows) in a data set, association-rule mining focuses on looking at relationships between attributes (or columns) in a data set. In a general sense, it looks for correlations—measured as co-occurrences—between products. Using association-rule mining, a business can start to answer questions about its customers' behaviors by looking for patterns that may exist in the data. Questions that market-basket analysis can be used to answer include: *Did a marketing campaign work? Have this customer's buying patterns changed? Has the customer had a major life event? Does the product location affect buying behavior? Who should we target with our new product?*

The Apriori algorithm is the main algorithm used to produce the association rules. It has a two-step process:

1. Find all combinations of items in a set of transactions that occur with a specified minimum frequency. These combinations are called *frequent itemsets*.

2. Generate rules that express the probable co-occurrence of items within frequent itemsets. The Apriori algorithm calculates the probability of an item being present in a frequent itemset given that another item or items are present.

The Apriori algorithm generates association rules that express probabilistic relationships between items in frequent itemsets. An association rule is of the form "IF *antecedent*, THEN *consequent*." It states that an item or group of items, the antecedent, implies the presence of another item in the same basket of goods, the consequent, with some probability. For example, a rule derived from a frequent itemset containing *A*, *B*, and *C* might state that if *A* and *B* are included in a transaction, then *C* is likely to also be included:

IF {hot-dogs, ketchup}, THEN {beer}.

This rule indicates that customers who are buying *hot dogs* and *ketchup* are also likely to buy *beer*. A frequent example of the power of association-rule mining is the *beer-diapers* example that describes how an unknown US supermarket in the 1980s used an early computer system to analyze its checkout data and identified an unusual association between diapers and beer in customer purchases. The theory developed to understand this rule was that families with young children were preparing for the weekend and knew that they would need diapers and would have to socialize at home. The store placed the two items near each other, and sales soared. The beer-and-diapers story has been debunked as apocryphal, but it is still a useful example of the potential benefits of association-rule mining for retail businesses.

Two main statistical measures are linked with association rules: *support* and *confidence*. The *support* percentage of an association rule—or the ratio of transactions that include both the antecedent and consequent to the total number of transactions—indicates how frequently the items in the rule occur together. The *confidence* percentage of an association rule—or the ratio of the number of transactions that include both the antecedent and consequent to the number of transactions that includes the antecedent—is the conditional probability that the consequent will occur given the occurrence of the antecedent. So, for example, a confidence of 75 percent for the association rule relating *hot dogs* and *ketchup* with *beer* would indicate that in 75 percent of cases where customers purchased both *hot dogs* and *ketchup*, they also purchased *beer*. The support score of a rule simply records the percentage of baskets in the data set where the rule holds. For example, a support of 5 percent indicates that 5 percent of all the baskets in the data set contain all three items in the rule "*hot dogs*, *ketchup*, and *beer*."

Even a small data set can result in the generation of a large number of association rules. In order to control the complexity of the analysis of these rules, it is usual to prune the generated rule set to include only rules that have both a high support and a high confidence. Rules that don't have high support or confidence are not interesting either because the rule covers only a very small percentage of baskets (low support) or because the relationship between the items in the antecedent and the consequent is low (low confidence). Rules that are trivial or inexplicable should also be pruned. Trivial rules represent associations that are obvious and well known to anyone who understands the business domain. An inexplicable rule represents associations that are so strange that it is difficult to understand how to convert the rule into a useful action for the company. It is likely that an inexplicable rule is the result of an odd data sample (i.e., the rule represents a spurious correlation). Once the rule set has been pruned, the data scientist can then analyze the remaining rules to understand what products are associated with each other and apply this new information in the organization. Organizations will typically use this new information to determine store layout or to perform some targeted marketing campaigns to their customers. These campaigns can involve updates to their websites to include recommended products, in-store advertisements, direct mailings, the cross-selling of other products by check-out staff, and so on.

Association mining becomes more powerful if the baskets of items are connected to demographic data about the customer. This is why so many retailers run loyalty-card schemes because such schemes allow them not only to connect different baskets of goods to the same customer through time but also to connect baskets of goods to the customer's demographics. Including this demographic information in the association analysis enables the analysis to be focused on particular demographics, which can further help marketing and targeted advertising. For example, demographic-based association rules can be used with new customers, for whom the company has no buying-habit information but does have demographic information. An example of an association rule augmented with demographic information might be

IF gender(male) and age(< 35) and {hot-dogs, ketchup},
THEN {beer}.

[Support = 2%, Confidence = 90%.]

The standard application area for association-rule mining focuses on what products are in the shopping basket and what products are not in the shopping basket. This assumes that the products are purchased in one visit to the store or website. This kind of scenario will probably work in most retail and other related scenarios. However, association-rule mining is also useful in a range of domains outside of retail. For example, in the telecommunications industry, applying association-rule mining to customer usage helps telecommunications companies to design how to bundle different services together into packages. In the insurance industry, association-rule mining is used to see if there are associations between products and claims. In the medical domain, it is used to check if there are interactions between existing and new treatments and medicines. And in banking and financial services, it is used to see what products customers typically have and whether these products can be applied to new or existing customers. Association-rule mining can also be used to analyze purchasing behavior over a period of time. For example, customers tend to buy product X and Y today, and in three months' time they buy product Z. This time period can be considered a shopping basket, although it is one that spans three months. Applying association-rule mining to this kind of temporally defined basket expands the applications areas of association-rule mining to include maintenance schedules, the replacement of parts, service calls, financial products, and so on.

Churn or No Churn, That Is the Question (Classification)
A standard business task in customer-relationship management is to estimate the likelihood that an individual customer will take an action. The term *propensity modeling*

is used to describe this task because the goal is to model an individual's propensity to do something. This action could be anything from responding to marketing to defaulting on a loan or leaving a service. The ability to identify customers who are likely to leave a service is particularly important to cell phone service companies. It costs a cell phone service company a substantial amount of money to attract new customers. In fact, it is estimated that it generally costs five to six times more to attract a new customer than it does to retain an established one (Verbeke et al. 2011). As a result, many cell phone service companies are very keen to retain their current customers. However, they also want to minimize costs. So although it would be easy to retain customers by simply giving all customers reduced rates and great phone upgrades, this is not a realistic option. Instead, they want to target the offers they give their customers to just those customers who are likely to leave in the near future. If they can identify a customer who is about to leave a service and persuade that customer to stay, perhaps by offering her an upgrade or a new billing package, then they can save the difference between the price of the enticement they gave the customer and the cost of attracting a new customer.

The term *customer churn* is used to describe the process of customers leaving one service and joining another. So the problem of predicting which customers are likely to leave in the near future is known as *churn prediction*. As the name suggests, this is a prediction task. The prediction task is to classify a customer as being a churn risk or not. Many companies are using this kind of analysis to predict churn customers in the telecommunications, utilities, banking, insurance, and other industries. A growing area that companies are focusing on is the prediction of staff turnover or staff churn: which staff are likely to leave the company within a certain time period.

When a prediction model returns a label or category for an input, it is known as a *classification model*. Training a classification model requires historic data, where each instance is labeled to indicate whether the target event has happened for that instance. For example, customer-churn classification requires a data set in which each customer (one row per customer) is assigned a label indicating whether he or she has churned. The data set will include an attribute, known as the *target attribute*, that lists this label for each customer. In some instances, assigning a churn label to a customer record is a relatively straightforward task. For example, the customer may have contacted the organization and explicitly canceled his subscription or contract. However, in other cases the churn event may not be explicitly signaled. For example, not all cell phone customers have a monthly contract. Some customers have a pay-as-you-go (or prepay) contract in which they top up their account at irregular intervals when they need more phone credit. Defining whether a customer with this type of contract has churned can be difficult: Has a customer who hasn't made a call in two weeks churned, or is it necessary for a customer to have a zero balance and no activity for three weeks before she is considered to have churned? Once the churn event has been defined from a business perspective, it is then necessary to implement this definition in code in order to assign a target label to each customer in the data set.

Another complicating factor in constructing the training data set for a churn-prediction model is that time lags need to be taken into account. The goal of churn prediction is to model the propensity (or likelihood) that a customer will churn at some point in the future. As a consequence, this type of model has a temporal dimension that needs to be considered during the creation of the data set. The set of attributes in a propensity-model data set are drawn from two separate time periods: the *observation period* and the *outcome period*. The observation period is when the values of the input attributes are calculated. The outcome period is when the target attribute is calculated. The business goal of creating a customer-churn model is to enable the business to carry out some sort of intervention before the customer churns—in other words, to entice the customer to stay with the service. This means that the prediction about the customer churning must be made sometime in advance of the customer's actually leaving the service. The length of this period is the length of the outcome period, and the prediction that the churn model returns is actually that a customer will churn within this outcome period. For example, the model might be trained to predict that the customer will churn within one month or two months, depending on the speed of the business process to carry out the intervention.

Defining the outcome period affects what data should be used as input to the model. If the model is designed to predict that a customer will churn within two months from the day the model is run on that customer's record, then when the model is being trained, the input attributes that describe the historic customers who have already churned should be calculated using only the data that were available about those customers two months prior to their leaving the service. The input attributes describing currently active customers should similarly be calculated with the data available about these customers' activity two months earlier. Creating the data set in this way ensures that all the instances in the data set, including both churned and active customers, describe the customers at the time in their individual customer journeys that the model is being designed to make a prediction about them: in this example, two months *before* they churn or stay.

Nearly all customer-propensity models will use attributes describing the customer's demographic information as input: *age*, *gender*, *occupation*, and so on. In scenarios relating to an ongoing service, they are also likely to include attributes describing the customer's position in the customer life cycle: *coming on board*, *standing still midcycle*, *approaching end of a contract*. There are also likely to be attributes that are specific to the industry. For example, typical attributes used in telecommunication industry customer-churn models include the customer's average bill, changes in billing amount, average usage, staying within or generally exceeding plan minutes, the ratio of calls within the network to those outside the network, and potentially the type of phone used.[1] However, the specific attributes used in each model will vary from one project to the next. Gordon Linoff and Michael Berry (2011) report that in one churn-prediction project in South Korea, the researchers found it useful to include an attribute that described the churn rate associated with a customer's phone (i.e., What percentage of customers with this particular phone churned during the observation period?). However, when they went to build a similar customer-churn model in Canada, the handset/churn-rate attribute was useless. The difference was that in South Korea the cell phone service company offered large discounts on new phones to new customers, whereas in Canada the same discounts were offered to both existing and new customers. The overall effect was that in South Korea phones going out of date drove customer churn; people were incentivized to leave one operator for another in order to avail themselves of discounts, but in Canada this incentive to leave did not exist.

Once a labeled data set has been created, the major stage in creating a classification model is to use an ML algorithm to build the classification model. During modeling, it is good practice to experiment with a number of different ML algorithms to find out which algorithm works best on the data set. Once the final model has been selected, the likely accuracy of the predictions of this model on new instances is estimated by testing it on a subset of the data set that was not used during the model-training phase. If a model is deemed accurate enough and suitable for the business need, the model is then deployed and applied to new data either in a batch process or in real time. A really important part of deploying the model is ensuring that the appropriate business processes and resources are put in place so that the model is used effectively. There is no point in creating a customer-churn model unless there is a process whereby the model's predictions result in triggering customer interventions so that the business retains customers.

In addition to predicting the classification label, prediction models can also give a measure of how confident the model is in the prediction. This measure is called the *prediction probability* and will have a value between 0 and 1. The higher the value, the more likely the prediction is correct. The prediction-probability value can be used to prioritize which customers to focus on. For example, in customer-churn prediction the organization wants to concentrate on the customers who are most likely to leave. By using the prediction probability and sorting the churners based on this value, a business can focus on the key customers (those most likely to leave) first before moving on to customers with a lower prediction-probability score.

How Much Will It Cost? (Regression)

Price prediction is the task of estimating the price that a product will cost at a particular point in time. The product could be a car, a house, a barrel of oil, a stock, or a medical procedure. Having a good estimate of what something will cost is obviously valuable to anyone who is considering buying the item. The accuracy of a price-prediction model is domain dependent. For example, due to the variability in the stock market, predicting the price of a stock tomorrow is very difficult. By comparison, it may be easier to predict the price of a house at an auction because the variation in house prices fluctuates much more slowly than stocks.

The fact that price prediction involves estimating the value of a continuous attribute means that it is treated as a *regression problem*. A regression problem is structurally very similar to a classification problem; in both cases, the data science solution involves building a model that can predict the missing value of an attribute given a set of input attributes. The only difference is that classification involves estimating the value of a categorical attribute and regression involves estimating the value of a continuous attribute. Regression analysis requires a data set where the value of the target attribute for each of the historic instances is listed. The multi-input linear-regression model introduced in chapter 4 illustrated the basic structure of a regression model, with most other regression models being variants of this approach. The basic structure of a regression model for price prediction is the same no matter what product it is applied to; all that varies are the name and number of the attributes. For example, to predict the price of a house, the input would include attributes such as the size of the house, the number of rooms, the number of floors, the average house price in the area, the average house size in the area, and so on. By comparison, to predict the price of a car, the attributes would include the age of the car, the number of miles on the odometer, the engine size, the make of the car, the number of doors, and so on. In each case, given the appropriate data, the regression algorithm works out how each of the attributes contributes to the final price.

As has been the case with all the examples given throughout this chapter, the application example of using a regression model for price prediction is illustrative only of the type of problem that it is appropriate to frame as a regression-modeling task. Regression prediction can be used in a wide variety of other real-world problems. Typical regression-prediction problems include calculating profit, value and volume of sales, sizes, demand, distances, and dosage.

PRIVACY AND ETHICS

The biggest unknown facing data science today is how societies will choose to answer a new version of the old question regarding how best to balance the freedoms and privacy of individuals and minorities against the security and interests of society. In the context of data science, this old question is framed as follows: What do we as a society view are reasonable ways to gather and use the data relating to individuals in contexts as diverse as fighting terrorism, improving medicine, supporting public-policy research, fighting crime, detecting fraud, assessing credit risk, providing insurance underwriting, and advertising to targeted groups?

The promise of data science is that it provides a way to understand the world through data. In the current era of big data, this promise is very tantalizing, and, indeed, a number of arguments can be used to support the development and adoption of data-driven infrastructure and technologies. One common argument relates to improving efficiency, effectiveness, and competiveness—an argument that, at least in the business context, is backed by some academic research. For example, a study involving 179 large publicly traded firms in 2011 showed that the more data driven a firm's decision making is, the more productive the firm is: "We find that firms that adopt DDD [data-driven decision making] have output and productivity that is 5–6% higher than what would be expected given their other investments and information technology usage" (Brynjolfsson, Hitt, and Kim 2011, 1).

Another argument for increased adoption of data science technologies and practices relates to securitization. For a long time, governments have used the argument that surveillance improves security. And since the terrorist attacks in the United States on September 11, 2001, as well as with each subsequent terrorist attack throughout the world, the argument has gained traction. Indeed, it was frequently used in the public debate caused by Edward Snowden's revelations about the US National Security Agency's PRISM surveillance program and the data it routinely gathered on US citizens. A stark example of the power of this argument is the agency's US$1.7 billion investment in a data center in Bluffdale, Utah, that has the ability to store huge amounts of intercepted communications (Carroll 2013).

At the same time, however, societies, governments, and business are struggling to understand the long-term implications of data science in a big-data world. Given the rapid development of technologies around data gathering, data storage, and data analysis, it is not surprising that the legal frameworks in place and the broader ethical discussions around data, in particular the question of individual privacy, are running behind these advances. Notwithstanding this difficulty, basic legal principles around data collection and usage are important to understand and are nearly always applicable. Also, the ethical debate around data usage and privacy has highlighted some worrying trends that we as individuals and citizens should be aware of.

Commercial Interests versus Individual Privacy

Data science can be framed as making the world a more prosperous and secure place to live. But these same arguments can be used by very different organizations that have very distinct agendas. For example, contrast calls by civil liberties groups for government to be more open and transparent in the gathering, use, and availability of data in the hope of empowering citizens to hold these same governments to account with similar calls from business communities who hope to use these data to increase their profits (Kitchin 2014a). In truth, data science is a double-edged sword. It can be used to improve our lives through more efficient government, improved medicine and health care, less-expensive insurance, smarter cities, reduced crime, and many more ways. At the same time, however, it can also be used to spy on our private lives, to target us with unwanted advertising, and to control our behavior both overtly and covertly (the fear of surveillance can affect us as much as the surveillance itself does).

The contradictory aspects of data science can often be apparent in the same applications. For example, the use of data science in health insurance underwriting uses third-party marketing data sets that contain information such as purchasing habits, web search history, along with hundreds of other attributes relating to people's lifestyles (Batty, Tripathi, Kroll, et al. 2010). The use of these third-party data is troublesome because it may trigger self-disciplining, wherein people avoid certain activities, such as visiting extreme-sports websites, for fear of incurring higher insurance premiums (Mayer-Schönberger and Cukier 2014). However, the justification for the use of these data is that it acts as a proxy for more invasive and expensive information sources, such as blood tests, and in the long term will reduce costs and premiums and thereby increase the number of people with health insurance (Batty, Tripathi, Kroll, et al. 2010).

The fault lines in the debate between the commercial benefits and ethical considerations of using data science are apparent in the discussions around the use of personal data for targeted marketing. From a business advertising perspective, the incentive to use personal data is that there is a relationship between personalizing marketing, services, and products, on the one hand, and the effectiveness of the marketing, on the other. It has been shown that the use of personal social network data—such as identifying consumers who are connected to prior customers—increases the effectiveness of a direct-mail marketing campaign for a telecommunications service by three to five times compared to traditional marketing approaches (Hill, Provost, and Volinsky 2006). Similar claims have been made about the effectiveness of data-driven personalization of online marketing. For example, a study of online cost and effectiveness of online targeted advertising in the United States in 2010 compared *run-of-the-network* marketing (when an advertising campaign is pushed out across a range of websites without specific targeting of users or sites) with *behavioral targeting*[1] (Beales 2010). The study found that behavioral marketing was both more expensive (2.68 times more) but also more effective, with a conversion rate more than twice that of run-of-the-network marketing. Another well-known study on the effectiveness of data-driven online advertising was conducted by researchers from the University of Toronto and MIT (Goldfarb and Tucker 2011). They used the enactment of a privacy-protection bill in the European Union (EU)[2] that limited the ability of advertising companies to track users' online behavior in order to compare the effectiveness of online advertising under the new restrictions (i.e., in the EU) and the effectiveness online advertising not under the new restrictions (i.e., in the United States and other non-EU countries). The study found that online advertising was significantly less effective under the new restrictions, with a reported drop of 65 percent in study participants' recorded purchasing intent. The results of this study have been contested (see, for example, Mayer and Mitchell 2012), but the study has been used to support the argument that the more data that are available about an individual, the more effective the advertising that is directed to that individual will be. Proponents of data-driven targeted marketing frame this argument as a win–win for both the advertiser and the consumer, claiming that advertisers lower marketing costs by reducing wasted advertising and achieve better conversions rates, and consumers get more relevant advertising.

This utopian perspective on the use of personal data for targeted marketing is at best based on a selective understanding of the problem. Probably one of the most worrying stories related to targeted advertising was reported in the *New York Times* in 2012 and involves the American discount retail store Target (Duhigg 2012). It is well known in marketing that one of the times in a person's life when his or her shopping habits change radically is at the conception and birth of a child. Because of this radical change, marketers see pregnancy as an opportunity to shift a person's shopping habits and brand loyalties, and many retailers use publicly available birth records to trigger personalized marketing for new parents, sending them offers relating to baby products. In order to get a competitive advantage, Target wanted to identify pregnant customers at an early stage (ideally during the second trimester) without the mother-to-be voluntarily telling Target that she was pregnant.[3] This insight would enable Target to begin its personalized marketing before other retailers knew the baby was on the way. To achieve this goal, Target initiated a data science project with the aim of predicting whether a customer was pregnant based on an analysis of her shopping habits. The starting point for the project was to analyze the shopping habits of women who had signed up for Target's baby-shower registry. The analysis revealed that expectant mothers tended to purchase larger quantities of unscented lotion at the beginning of the second trimester as well as certain dietary supplements throughout the first 20 weeks of pregnancy. Based on this analysis, Target created a data-driven model that used around 25 products and indictors and assigned each customer a "pregnancy-prediction" score. The *success*, for want of a better word, of this model was made very apparent when a man turned up at a Target store to complain about the fact that his high-school-age daughter had been mailed coupons for baby clothes and cribs. He accused Target of trying to encourage his daughter to get pregnant. However, over the subsequent days it transpired that the man's daughter was in fact pregnant but hadn't told anyone. Target's pregnancy-prediction model was able to identify a pregnant high school student and act on this information before she had chosen to tell her family.

Ethical Implications of Data Science: Profiling and Discrimination

The story about Target identifying a pregnant high school student without her consent or knowledge highlights how data science can be used for social profiling not only of individuals but also of minority groups in society. In his book *The Daily You: How the New Advertising Industry Is Defining Your Identity and Your Worth* (2013), Joseph Turow discusses how marketers use digital profiling to categorize people as either *targets* or *waste* and then use these categories to personalize the offers and promotions directed to individual consumers: "those considered waste are ignored or shunted to other products that marketers deem more relevant to their tastes or income" (11). This personalization can result in preferential treatment for some and

Personalization can result in preferential treatment for some and marginalization of others.

marginalization of others. A clear example of this discrimination is differential pricing on websites, wherein some customers are charged more than other customers for the same product based on their customer profiles (Clifford 2012).

These profiles are constructed by integrating data from a number of different noisy and partial data sources, so the profiles can often be misleading about an individual. What is worse is that these marketing profiles are treated as products and are often sold to other companies, with the result that a negative marketing assessment of an individual can follow that individual across many domains. We have already discussed the use of marketing data sets in insurance underwriting (Batty, Tripathi, Kroll, et al. 2010), but these profiles can also make their way into credit-risk assessments and many other decision processes that affect people's lives. Two aspects of these marketing profiles make them particularly problematic. First, they are a black box, and, second, they are persistent. The black-box nature of these profiles is apparent when one considers that it is difficult for an individual to know what data are recorded about them, where and when the data were recorded, and how the decision processes that use these data work. As a result, if an individual ends up on a no-fly list or a credit blacklist, it is "difficult to determine the grounds for discrimination and to challenge them" (Kitchin 2014a, 177). What is more, in the modern world data are often stored for a long time. So data recorded about an event in an individual's life persists long after an event. As Turow warns, "Turning individual profiles into individual evaluations is what happens when a profile becomes a reputation" (2013, 6).

Furthermore, unless used very carefully, data science can actually perpetuate and reinforce prejudice. An argument is sometimes made that data science is objective: it is based on numbers, so it doesn't encode or have the prejudicial views that affect human decisions. The truth is that data science algorithms perform in an amoral manner more than in an objective manner. Data science extracts patterns in data; however, if the data encode a prejudicial relationship in society, then the algorithm is likely to identify this pattern and base its outputs on the pattern. Indeed, the more consistent a prejudice is in a society, the stronger that prejudicial pattern will appear in the data about that society, and the more likely a data science algorithm will extract and replicate that pattern of prejudice. For example, a study carried out by academic researchers on the Google Online Advertising system found that the system showed an ad relating to a high-paying job more frequently to participants whose Google profile identified them as male compared to participants whose profile identified them as female (Datta, Tschantz, and Datta 2015).

The fact that data science algorithms can reinforce prejudice is particularly troublesome when data science is applied to policing. Predictive Policing, or PredPol,[4] is a data science tool designed to predict when and where a crime is most likely to occur. When deployed in a city, PredPol generates a daily report listing a number of hot spots on a map (small areas 500 feet by 500 feet) where the system believes crimes are likely to occur and tags each hot spot with the police shift during which the system believes the crime will occur. Police departments in both the United States and the United Kingdom have deployed PredPol. The idea behind this type of intelligent-policing system is that policing resources can be efficiently deployed. On the surface, this seems like a sensible application of data science, potentially resulting in efficient targeting of crime and reducing policing costs. However, questions have been raised about the accuracy of PredPol and the effectiveness of similar predictive-policing initiatives (Hunt, Saunders, and Hollywood 2014; Oakland Privacy Working Group 2015; Harkness 2016). The potential for these types of systems to encode racial or class-based profiling in policing has also been noted (Baldridge 2015). The deployment of police resources based on historic data can result in a higher police presence in certain areas—typically economically disadvantaged areas—which in turn results in higher levels of reported crime in these areas. In other words, the prediction of crime becomes a self-fulfilling prophesy. The result of this cycle is that some locations will be disproportionately targeted by police surveillance, causing a breakdown in trust between the people who live

Unless used very carefully, data science can actually perpetuate and reinforce prejudice.

in those communities and policing institutions (Harkness 2016).

Another example of data-driven policing is the Strategic Subjects List (SSL) used by the Chicago Police Department in an attempt to reduce gun crime. The list was first created in 2013, and at that time it listed 426 people who were estimated to be at a very high risk of gun violence. In an attempt to proactively prevent gun crime, the Chicago Police Department contacted all the people on the SSL to warn them that they were under surveillance. Some of the people on the list were very surprised to be included on it because although they did have criminal records for minor offenses, they had no violence on their records (Gorner 2013). One question to ask about this type of data gathering to prevent crime is, How accurate is the technology? A recent study found that the people on the SSL for 2013 were "not more or less likely to become a victim of a homicide or shooting than the comparison group" (Saunders, Hunt, and Hollywood 2016). However, this study also found that individuals on the list were more likely to be arrested for a shooting incident, although it did point out that this greater likelihood could have been created by the fact that these individuals were on the list, which resulted in increasing police officers' awareness of these individuals (Saunders, Hunt, and Hollywood 2016). Responding to this study, the Chicago Police Department stated that it regularly updated the algorithm used to compile the SSL and that the effectiveness of the SSL had improved since 2013 (Rhee 2016). Another question about data-driven crime-prevention lists is, How does an individual end up on the list? The 2013 version of the SSL appears to have been compiled using, among other attributes of an individual, an analysis of his or her social network, including the arrest and shooting histories of his or her acquaintances (Dokoupil 2013; Gorner 2013). On the one hand, the idea of using social network analysis makes sense, but it opens up the very real problem of guilt by association. One problem with this type of approach is that it can be difficult to define precisely what an association between two individuals entails. Is living on the same street enough to be an association? Furthermore, in the United States, where the vast majority of inmates in prison are African American and Latino males, allowing predictive-policing algorithms to use the concept of association as an input is likely to result in predictions targeting mainly young men of color (Baldridge 2015).

The anticipatory nature of predictive policing means that individuals may be treated differently not because of what they have done but because of data-driven inferences about what they might do. As a result, these types of systems may reinforce discriminatory practices by replicating the patterns in historic data and may create self-fulfilling prophecies.

Ethical Implications of Data Science: Creating a Panopticon

If you spend time absorbing some of the commercial boosterism that surrounds data science, you get a sense that any problem can be solved using data science technology given enough of the right data. This marketing of the power of data science feeds into a view that a data-driven approach to governance is the best way to address complex social problems, such as crime, poverty, poor education, and poor public health: all we need to do to solve these problems is to put sensors into our societies to track everything, merge all the data, and run the algorithms to generate the key insights that provide the solution.

When this argument is accepted, two processes are often intensified. The first is that society becomes more technocratic in nature, and aspects of life begin to be regulated by data-driven systems. Examples of this type of technological regulation already exist—for example, in some jurisdictions data science is currently used in parole hearings (Berk and Bleich 2013) and sentencing (Barry-Jester, Casselman, and Goldstein 2015). For an example outside of the judicial system, consider how smart-city technologies regulate traffic flows through cities with algorithms dynamically deciding which traffic flow gets priority at a junction at different times of day (Kitchin 2014b). A by-product of this technocratic regulation is the proliferation of the sensors that support the automated regulating systems. The second process is "control creep," wherein data gathered for one purpose is repurposed and used to regulate in another way (Innes 2001). For example, road cameras that were installed in London with the primary purpose of regulating congestion and implementing congestion charges (the London congestion charge is a daily charge for driving a vehicle within London during peak times) have been repurposed for security tasks (Dodge and Kitchin 2007). Other examples of control creep include a technology called ShotSpotter that consists of a city-wide network of microphones designed to identify gunshots and report the locations of them but that also records conversations, some of which were used to achieve criminal convictions (Weissman 2015), and the use of in-car navigation systems to monitor and fine rental car drivers who drive out of state (Elliott 2004; Kitchin 2014a).

An aspect of control creep is the drive to merge data from different sources so as to provide a more complete picture of a society and thereby potentially unlock deeper insights into the problems in the system. There are often good reasons for the repurposing of data. Indeed, calls are frequently made for data held by different branches of government to be merged for legitimate purposes—for example, to support health research and for the convenience of the state and its citizens. From a civil liberties perspective, however, these trends are very concerning.

Heightened surveillance, the integration of data from multiple sources, control creep, and anticipatory governance (such as the predictive-policing programs) may result in a society where an individual may be treated with suspicion simply because a sequence of unrelated innocent actions or encounters matches a pattern deemed suspicious by a data-driven regulatory system. Living in this type of a society would change each of us from free citizens into inmates in Bentham's Panopticon,[5] constantly self-disciplining our behaviors for fear of what inferences may be drawn from them. The distinction between individuals who believe and act as though they are free of surveillance and individuals who self-discipline out of fear that they inhabit a Panopticon is the primary difference between a free society and a totalitarian state.

Á la recherche du privacy perdu

As individuals engage with and move through technically modern societies, they have no choice but to leave a data trail behind them. In the real world, the proliferation of video surveillance means that location data can be gathered about an individual whenever she appears on a street or in a shop or car park, and the proliferation of cell phones means that many people can be tracked via their phones. Other examples of real-world data gathering include the recording of credit card purchases, the use of loyalty schemes in supermarkets, the tracking of withdrawals from ATMs, and the tracking of cell phone calls made. In the online world, data are gathered about individuals when they visit or log in to websites; send an email; engage in online shopping; rate a date, restaurant, or store; use an e-book reader; watch a lecture in a massive open online course; or like or post something on a social media site. To put into perspective the amount of data that are gathered on the average individual in a technologically modern society, a report from the Dutch Data Protection Authority in 2009 estimated that the average Dutch citizen was included in 250 to 500 databases, with this figure rising to 1,000 databases for more socially active people (Koops 2011). Taken together, the data points relating to an individual define that person's *digital footprint*.

The data in a digital footprint can be gathered in two contexts that are problematic from a privacy perspective. First, data can be collected about an individual without his knowledge or awareness. Second, in some contexts an individual may choose to share data about himself and his opinions but may have little or no knowledge of or control over how these data are used or how they will be shared with and repurposed by third parties. The terms *data shadow* and *data footprint*[6] are used to distinguish these two contexts of data gathering: an individual's data shadow comprises the data gathered about an individual without her knowledge, consent, or awareness, and an individual's data footprint consists of the pieces of data that she knowingly makes public (Koops 2011).

The collection of data about an individual without her knowledge or consent is of course worrying. However, the power of modern data science techniques to uncover hidden patterns in data coupled with the integration and repurposing of data from several sources means that even data collected with an individual's knowledge and consent in one context can have negative effects on that individual that are impossible for them to predict. Today, with the use of modern data science techniques, very personal information that we may not want to be made public and choose not to share can still be reliably inferred from seemingly unrelated data we willingly post on social media. For example, many people are willing to like something on Facebook because they want to demonstrate support to a friend. However, by simply using the items that an individual has liked on Facebook, data-driven models can accurately predict that person's sexual orientation, political and religious views, intelligence and personality traits, and use of addictive substances such as alcohol, drugs, and cigarettes; they can even determine whether that person's parents stayed together until he or she was 21 years old (Kosinski, Stillwell, and Graepel 2013). The out-of-context linkages made in these models is demonstrated by how liking a human rights campaign was found to be predictive of homosexuality (both male and female) and by how liking Hondas was found to be predictive of not smoking (Kosinski, Stillwell, and Graepel 2013).

Computational Approaches to Preserving Privacy

In recent years, there has been a growing interest in computational approaches to preserving individual privacy throughout a data-analysis process. Two of the best-known approaches are *differential privacy* and *federated learning*.

Differential privacy is a mathematical approach to the problem of learning useful information about a population while at the same time learning nothing about the individuals within the population. Differential privacy uses a particular definition of privacy: the privacy of an individual has not been compromised by the inclusion of his or her data in the data-analysis process if the conclusions reached by the analysis would have been the same independent of whether the individual's data were included or not. A number of processes can be used to implement differential privacy. At the core of these processes is the idea of injecting noise either into the data-collection process or into the responses to database queries. The noise protects the privacy of individuals but can be removed from the data at an aggregate level so that useful population-level statistics can be calculated. A useful example of a procedure for injecting noise into data that provides an intuitive explanation of how differential privacy processes can work is the randomized-response technique. The use case for this technique is a survey that includes a sensitive yes/no question (e.g., relating to law breaking, health conditions, etc.). Survey respondents are instructed to answer the sensitive question using the following procedure:

1. Flip a coin and keep the result of the coin flip secret.

2. If tails, respond "Yes."

3. If heads, respond truthfully.

Half the respondents will get tails and respond "Yes"; the other half will respond truthfully. Therefore, the true number of "No" respondents in the total population is (approximately) twice the number of "No" responses (the coin is fair and selects randomly, so the distribution of yes/no responses among the respondents who got tails should mirror the number of respondents who answered truthfully). Given the true count for "No," we can calculate the true count for "Yes." However, although we now have an accurate count for the population regarding the sensitive "Yes" condition, it is not possible to identify for which of the "Yes" respondents the sensitive condition actually holds. There is a trade-off between the amount of noise injected into data and the usefulness of the data for data analysis. Differential privacy addresses this trade-off by providing estimates of the amount of noise required given factors such as the distribution of data within the database, the type of database query that is being processed, and the number of queries through which we wish to guarantee an individual's privacy. Cynthia Dwork and Aaron Roth (2014) provide an introduction to differential privacy and an overview of several approaches to implementing differential privacy. Differential-privacy techniques are now being deployed in a number of consumer products. For example, Apple uses differential privacy in iOS 10 to protect the privacy of individual users while at the same time learning usage patterns to improve predictive text in the messaging application and to improve search functionality.

In some scenarios, the data being used in a data science project are coming from multiple disparate sources. For example, multiple hospitals may be contributing to a single research project, or a company is collecting data from a large number of users of a cell phone application. Rather than centralizing these data into a single data repository and doing the analysis on the combined data, an alternative approach is to train different models on the subsets of the data at the different data sources (i.e., at the individual hospitals or on the phones of each individual user) and then to merge the separately trained models. Google uses this federated-learning approach to improve

The truth is that data science algorithms perform in an amoral manner more than in an objective manner.

the query suggestions made by the Google keyboard on Android (McMahan and Ramage 2017). In Google's federated-learning framework, the mobile device initially has a copy of the current application loaded. As the user uses the application, the application data for that user are collected on his phone and used by a learning algorithm that is local to the phone to update the local version of the model. This local update of the model is then uploaded to the cloud, where it is averaged with the model updates uploaded from other user phones. The core model is then updated using this average. With the use of this process, the core model can be improved, and individual users' privacy can at the same time be protected to the extent that only the model updates are shared—not the users' usage data.

Legal Frameworks for Regulating Data Use and Protecting Privacy

There is variation across jurisdictions in the laws relating to privacy protection and permissible data usage. However, two core pillars are present across most democratic jurisdictions: antidiscrimination legislation and personal-data-protection legislation.

In most jurisdictions, antidiscrimination legislation forbids discrimination based on any of the following grounds: disability, age, sex, race, ethnicity, nationality, sexual orientation, and religious or political opinion. In the United States, the Civil Rights Act of 1964[7] prohibits discrimination based on color, race, sex, religion, or nationality. Later legislation has extended this list; for example, the Americans with Disabilities Act of 1990[8] extended protection to people against discrimination based on disabilities. Similar legalization is in place in many other jurisdictions. For example, the Charter of Fundamental Rights of the European Union prohibits discrimination based on any grounds, including race, color, ethnic or social origin, genetic features, sex, age, birth, disability, sexual orientation, religion or belief, property, membership in a national minority, and political or any other opinion (Charter 2000).

A similar situation of variation and overlap exists with respect to privacy legislation across different jurisdictions. In the United States, the Fair Information Practice Principles (1973)[9] have provided the basis for much of the subsequent privacy legislation in that jurisdiction. In the EU, the Data Protection Directive (Council of the European Union and European Parliament 1995) is the basis for much of that jurisdiction's privacy legislation. The General Data Protection Regulations (Council of the European Union and European Parliament 2016) expand on the data protection principles in the Data Protection Directive and provide consistent and legally enforceable data protection regulations across all EU member states.

However, the most broadly accepted principles relating to personal privacy and data are the Guidelines on the Protection of Privacy and Transborder Flows of Personal Data published by the Organisation for Economic Co-operation and Development (OECD 1980). Within these guidelines, personal data are defined as records relating to an identifiable individual, known as the *data subject*. The guidelines define eight (overlapping) principles that are designed to protect a data subject's privacy:

1. Collection Limitation Principle: Personal data should only be obtained lawfully and with the knowledge and consent of the data subject.

2. Data Quality Principle: Any personal data that are collected should be relevant to the purpose for which they are used; they should be accurate, complete, and up to date.

3. Purpose Specification Principle: At or before the time that personal data are collected, the data subject should be informed of the purpose for which the data will be used. Furthermore, although changes of purpose are permissible, they should not be introduced arbitrarily (new purposes must be compatible with the original purpose) and should be specified to the data subject.

4. Use Limitation Principle: The use of personal data is limited to the purpose that the data subject has been informed of, and the data should not be disclosed to third

parties without the data subject's consent or by authority of law.

5. Safety Safeguards Principle: Personal data should be protected by security safeguards against deletion, theft, disclosure, modification, or unauthorized use.

6. Openness Principle: A data subject should be able to acquire information with reasonable ease regarding the collection, storage, and use of his or her personal data.

7. Individual Participation Principle: A data subject has the right to access and challenge personal data.

8. Accountability Principle: A data controller is accountable for complying with the principles.

Many countries, including the EU and the United States, endorse the OECD guidelines. Indeed, the data protection principles in the EU General Data Protection Regulations can be broadly traced back to the OECD guidelines. The General Data Protection Regulations apply to the collection, storage, transfer and processing of personal data relating to EU citizens within the EU and has implications for the flows of this data outside of the EU. Currently, several countries are developing data protection laws similar to and consistent with the General Data Protection Regulations.

Toward an Ethical Data Science

It is well known that despite the legal frameworks that are in place, nation-states frequently collect personal data on their citizens and foreign nationals without these people's knowledge, often in the name of security and intelligence. Examples include the US National Security Agency's PRISM program; the UK Government Communications Headquarters' Tempora program (Shubber 2013); and the Russian government's System for Operative Investigative Activities (Soldatov and Borogan 2012). These programs affect the public's perception of governments and use of modern communication technologies. The results of the Pew survey "Americans' Privacy Strategies Post-Snowden" in 2015 indicated that 87 percent of respondents were aware of government surveillance of phone and Internet communications, and among those who were aware of these programs 61 percent stated that they were losing confidence that these programs served the public interest, and 25 percent reported that they had changed how they used technologies in response to learning about these programs (Rainie and Madden 2015). Similar results have been reported in European surveys, with more than half of Europeans aware of large-scale data collection by government agencies and most respondents stating that this type of surveillance had a negative impact on their trust with respect to how their online personal data are used (Eurobarometer 2015).

At the same time, many private companies avoid the regulations around personal data and privacy by claiming to use derived, aggregated, or anonymized data. By repackaging data in these ways, companies claim that the data are no longer personal data, which, they argue, permits them to gather data without an individual's awareness or consent and without having a clear immediate purpose for the data; to hold the data for long periods of time; and to repurpose the data or sell the data when a commercial opportunity arises. Many advocates of the commercial opportunities of data science and big data argue that the real commercial value of data comes from their reuse or "optional value" (Mayer-Schönberger and Cukier 2014). The advocates of data reuse highlight two technical innovations that make data gathering and storage a sensible business strategy: first, today data can be gathered passively with little or no effort or awareness on the part of the individuals being tracked; and, second, data storage has become relatively inexpensive. In this context, it makes commercial sense to record and store data in case future (potentially unforeseeable) commercial opportunities make it valuable.

The modern commercial practices of hoarding, repurposing, and selling data are completely at odds with the purpose specification and use-limitation principles of the OECD guidelines. Furthermore, the collection-limitation principle is undermined whenever a company presents a privacy agreement to a consumer that is designed to be unreadable or reserves the right for the company to modify the agreement without further consultation or notification or both. Whenever this happens, the process of notification and granting of consent is turned into a meaningless box-ticking exercise. Similar to the public opinion about government surveillance in the name of security, public opinion is quite negative toward commercial websites' gathering and repurposing of personal data. Again using American and European surveys as our litmus test for wider public opinion, a survey of American Internet users in 2012 found that 62 percent of adults surveyed stated that they did not know how to limit the information collected about them by websites, and 68 percent stated that they did not like the practice of targeted advertising because they did not like their online behavior tracked and analyzed (Purcell, Brenner, and Rainie 2012). A recent survey of European citizens found similar results: 69 percent of respondents felt that the collection of their data should require their explicit approval, but only 18 percent of respondents actually fully read privacy statements. Furthermore, 67 percent of respondents stated that they don't read privacy statements because they found them too long, and 38 percent stated that they found them unclear or too difficult to understand. The survey also found that 69 percent of respondents were concerned about their information being used for different purposes from the one it was collected for, and 53 percent of respondents were uncomfortable with Internet companies using their personal information to tailor advertising (Eurobarometer 2015).

So at the moment public opinion is broadly negative toward both government surveillance and Internet companies' gathering, storing, and analyzing of personnel data. Today, most commentators agree that data-privacy legislation needs to be updated and that changes are happening. In 2012, both the EU and the United States published reviews and updates relating to data-protection and privacy policies (European Commission 2012; Federal Trade Commission 2012; Kitchin 2014a, 173). In 2013, the OECD guidelines were extended to include, among other updates, more details in relation to implementing the accountability principle. In particular, the new guidelines define the data controller's responsibilities to have a privacy-management program in place and to define clearly what such a program entails and how it should be framed in terms of risk management in relation to personal data (OECD 2013). In 2014, a Spanish citizen, Mario Costeja Gonzalez, won a case in the EU Court of Justice against Google (C-131/12 [2014]) asserting his right to be forgotten. The court held that an individual could request, under certain conditions, an Internet search engine to remove links to webpages that resulted from searches on the individual's name. The grounds for such a request included that the data are inaccurate or out of date or that the data had been kept for longer than was necessary for historical, statistical, or scientific purposes. This ruling has major implications for all Internet search engines but may also have implications for other big-data hoarders. For example, it is not clear at present what the implications are for social media sites such as Facebook and Twitter (Marr 2015). The concept of the right to be forgotten has been asserted in other jurisdictions. For example, the California "eraser" law asserts a minor's right to have material he has posted on an Internet or mobile service removed at his request. The law also prohibits Internet, online, or cell phone service companies from compiling personal data relating to a minor for the purposes of targeted advertising or allowing a third party to do so.[10] As a final example of the changes taking place, in 2016 the EU-US Privacy Shield was signed and adopted (European Commission 2016). Its focus is on harmonizing data-privacy obligations across the two jurisdictions. Its purpose is to strengthen the data-protection rights for EU citizens in the context where their data have been moved outside of the EU. This agreement imposed stronger obligations on commercial companies with regard to transparency of data usage, strong oversight mechanisms and possible sanctions, as well as limitations and oversight mechanisms for public authorities in recording or accessing personal data. However, at the time of writing, the strength and effectiveness of the EU-US Privacy Shield is being tested in a legal case in the Irish courts. The reason why the Irish legal system is at the center of this debate is that many of the large US multinational Internet companies (Google, Facebook, Twitter, etc.) have their European, Middle East, and Africa headquarters in Ireland. As a result, the data-protection commissioner for Ireland is responsible for enforcing EU regulations on transnational data transfers made by these companies. Recent history illustrates that it is possible for legal cases to result in significant and swift changes in the regulation of how personnel data are handled. In fact, the EU-US Privacy Shield is a direct consequence of a suit filed by Max Schrems, an Austrian lawyer and privacy activist, against Facebook. The outcome of Schrems's case in 2015 was to invalidate the existing EU-US Safe Harbor agreement with immediate effect, and the EU-US Privacy Shield was developed as an emergency response to this outcome. Compared to the original Safe Harbor agreement, the Privacy Shield has strengthened EU citizens' data-privacy rights (O'Rourke and Kerr 2017), and it may well be that any new framework would further strengthen these rights. For example, the EU General Data Protection Regulations will provide legally enforceable data protection to EU citizens from May 2018.

From a data science perspective, these examples illustrate that the regulations around data privacy and protection are in flux. Admittedly, the examples listed here are from the US and EU contexts, but they are indicative of broader trends in relation to privacy and data regulation.

It is very difficult to predict how these changes will play out in the long term. A range of vested interests exist in this domain: consider the differing agendas of big Internet, advertising and insurances companies, intelligence agencies, policing authorities, governments, medical and social science research, and civil liberties groups. Each of these different sectors of society has differing goals and needs with regard to data usage and consequently has different views on how data-privacy regulation should be shaped. Furthermore, we as individuals will probably have shifting views depending on the perspective we adopt. For example, we might be quite happy for our personnel data to be shared and reused in the context of medical research. However, as the public-opinion surveys in Europe and the United States have reported, many of us have reservations about data gathering, reuse, and sharing in the context of targeted advertising. Broadly speaking, there are two themes in the discourse around the future of data privacy. One view argues for the strengthening of regulations relating to the gathering of personal data and in some cases empowering individuals to control how their data are gathered, shared, and used. The other view argues for deregulation in relation to the gathering of data but also for stronger laws to redress the misuse of personnel data. With so many different stakeholders and perspectives, there are no easy or obvious answers to the questions posed about privacy and data. It is likely that the eventual solutions that are developed will be defined on a sector-by-sector basis and consist of compromises negotiated between the relevant stakeholders.

In such a fluid context, it is best to act conservatively and ethically. As we work on developing new data science solutions to business problems, we should consider ethical questions in relation to personal data. There are good business reasons to do so. First, acting ethically and transparently with personal data will ensure that a business will have good relationships with its customers. Inappropriate practices around personal data can cause a business severe reputational damage and cause its customer to move to competitors (Buytendijk and Heiser 2013). Second, there is a risk that as data integration, reuse, profiling, and targeting intensify, public opinion will harden around data privacy in the coming years, which will lead to more-stringent regulations. Consciously acting transparently and ethically is the best way to ensure that the data science solutions we develop do not run afoul of current regulations or of the regulations that may come into existence in the coming years.

Aphra Kerr (2017) reports a case from 2015 that illustrates how not taking ethical considerations into account can have serious consequences for technology developers and vendors. The case resulted in the US Federal Trade Commission fining app game developers and publishers under the Children's Online Privacy Protection Act. The developers had integrated third-party advertising into their free-to-play games. Integrating third-party advertising is standard practice in the free-to-play business model, but the problem arose because the games were designed for children younger than 13. As a result, in sharing their users' data with advertising networks, the developers where in fact also sharing data relating to children and as a result violated the Children's Online Privacy Protection Act. Also, in one instance the developers failed to inform the advertising networks that the apps were for children. As a result, it was possible that inappropriate advertising could be shown to children, and in this instance the Federal Trade Commission ruled that the game publishers were responsible for ensuring that age-appropriate content and advertising were supplied to the game-playing children. There has been an increasing number of these types of cases in recent years, and a number of organizations, including the Federal Trade Commission (2012), have called for businesses to adopt the principles of *privacy by design* (Cavoukian 2013). These principles were developed in the 1990s and have become a globally recognized framework for the protection of privacy. They advocate that protecting privacy should be the default mode of operation for the design of technology and information systems. To follow these principles requires a designer to consciously and proactively seek to embed privacy considerations into the design of technologies, organizational practices, and networked system architectures.

Although the arguments of ethical data science are clear, it is not always easy to act ethically. One way to make the challenge of ethical data science more concrete is to imagine you are working for a company as a data scientist on a business-critical project. In analyzing the data, you have identified a number of interacting attributes that together are a proxy for race (or some other personal attribute, such as religion, gender, etc.). You know that legally you can't use the race attribute in your model, but you believe that these proxy attributes would enable you to circumvent the antidiscrimination legislation. You also believe that including these attributes in the model will make your model work, although you are naturally concerned that this successful outcome may be because the model will learn to reinforce discrimination that is already present in the system. Ask yourself: "What do I do?"

FUTURE TRENDS AND PRINCIPLES OF SUCCESS

An obvious trend in modern societies is the proliferation of systems that can sense and react to the world: smart phones, smart homes, self-driving cars, and smart cities. This proliferation of smart devices and sensors presents challenges to our privacy, but it is also driving the growth of big data and the development of new technology paradigms, such as the Internet of Things. In this context, data science will have a growing impact across many areas of our lives. However, there are two areas where data science will lead to significant developments in the coming decade: personal medicine and the development of smart cities.

Medical Data Science

In recent years, the medical industry has been looking at and adopting data science and predictive analytics.

Doctors have traditionally had to rely on their experiences and instincts when diagnosing a condition or deciding on what the next treatment might be. The evidence-based medicine and precision-medicine movement argue that medical decisions should be based on data, ideally linking the best available data to an individual patient's predicament and preferences. For example, in the case of precision medicine, fast genome-sequencing technology means that it is now feasible to analyze the genomes of patients with rare diseases in order to identify mutations that cause the disease so as to design and select appropriate therapies specific to that individual. Another factor driving data science in medicine is the cost of health care. Data science, in particular predictive analytics, can be used to automate some health care processes. For example, predictive analytics has been used to decide when antibiotics and other medicines should be administrated to babies and adults, and it is widely reported that many lives have been saved because of this approach.

Medical sensors worn or ingested by the patient or implanted are being developed to continuously monitor a patient's vital signs and behaviors and how his or her organs are functioning throughout the day. These data are continuously gathered and fed back to a centralized monitoring server. It is here at the monitoring server that health care professionals access the data being generated by all the patients, assess their conditions, understand what effects the treatment is having, and compare each patient's results to those of other patients with similar conditions to inform them regarding what should happen next in each patient's treatment regime. Medical science is using the data generated by these sensors and integrating it with additional data from the various parts of the medical profession and the pharmaceutical industry to determine the effects of current and new medicines. Personalized treatment programs are being developed based on the type of patient, his condition, and how his body responds to various medicines. In addition, this new type of medical data science is now feeding into new research on medicines and their interactions, the design of more efficient and detailed monitoring systems, and the uncovering of greater insights from clinical trials.

Smart Cities

Various cities around the world are adopting new technology to be able to gather and use the data generated by their citizens in order to better manage the cities' organizations, utilities, and services. There are three core enablers of this trend: data science, big data, and the Internet of Things. The name "Internet of Things" describes the internetworking of physical devices and sensors so that these devices can share information. This may sound mundane, but it has the benefit that we can now remotely control smart devices (such as our home if it is properly configured) and opens the possibility that networked machine-to-machine communication will enable smart environments to autonomously predict and react to our needs (for example, there are now commercially available smart refrigerators that can warn you when food is about to spoil and allows you to order fresh milk through your smart phone).

Smart-city projects integrate real-time data from many different data sources into a single data hub, where they are analyzed and used to inform management and planning decisions. Some smart-city projects involve building brand-new cities that are smart from the ground up. Both Masdar City in the United Arab Emirates and Songdo City in South Korea are brand-new cities that have been built with the smart technology at their core and a focus on being eco-friendly and energy efficient. However, most smart-city projects involve the retrofitting of existing cities with new sensor networks and data-processing centers. For example, in the SmartSantander project in Spain,[1] more than 12,000 networked sensors have been installed across the city to measure temperature, noise, ambient lighting, carbon monoxide levels, and parking. Smart-city projects often focus on developing energy efficiency, planning and routing traffic, and planning utility services to match population needs and growth.

Japan has embraced the smart-city concept with a particular focus on reducing energy usage. The Tokyo Electric Power Company (TEPC) has installed more than 10 million smart meters across homes in the TEPC service area.[2] At the same time, TEPC is developing and rolling out smart-phone applications that enable customers to track the electricity used in their homes in real time and to change their electricity contract. These smart-phone applications also enable the TEPC to send each customer personalized energy-saving advice. Outside of the home, smart-city technology can be used to reduce energy usage through intelligent street lighting. The Glasgow Future Cities Demonstrator is piloting street lighting that switches on and off depending on whether people are present. Energy efficiency is also a top priority for all new buildings, particularly for large local government and commercial buildings. These buildings' energy efficiency can be optimized by automatically managing climate controls through a combination of sensor technology, big data, and data science. An extra benefit of these smart-building monitoring systems is that they can monitor for levels of pollution and air quality and can activate the necessary controls and warnings in real time.

Transport is another area where cities are using data science. Many cities have implemented traffic-monitoring and management systems. These systems use real-time data to control the flow of traffic through the city. For example, they can control traffic-light sequences in real time, in some cases to give priority to public-transport vehicles. Data on city transport networks are also useful for planning public transport. Cities are examining the routes, schedules, and vehicle management to ensure that services support the maximum number of people and to reduce the costs associated with delivering the transport services. In addition to modeling the public network, data science is also being used to monitor official city vehicles to ensure their optimal usage. Such projects combine traffic conditions (collected by sensors along the road network, at traffic lights, etc.), the type of task being performed, and other conditions to optimize route planning, and dynamic route adjustments are fed to the vehicles with live updates and changes to their routes.

Beyond energy usage and transport, data science is being used to improve the provision of utility services and to implement longer-term planning of infrastructure projects. The efficient provision of utility services is constantly being monitored based on current usage and projected usages, and the monitoring takes into account previous usage in similar conditions. Utility companies are using data science in a number of ways. One way is monitoring the delivery network for the utility: the supply, the quality of the supply, any network issues, areas that require higher-than-expected usage, automated rerouting of the supply, and any anomalies in the network. Another way that utility companies are using data science is in monitoring their customers. They are looking for unusual usage that might indicate some criminality (for example, a grow house), customers who may have altered the equipment and meters for the building where they live, and customers who are most likely to default on their payments. Data science is also being used in examining the best way to allocate housing and associated services in city planning. Models of population growth are built to forecast into the future, and based on various simulations the city planners can estimate when and where certain support services, such as high schools, are needed.

Data Science Project Principles: Why Projects Succeed or Fail

A data science project sometimes fails insofar as it doesn't deliver what was hoped for because it gets bogged down in some technical or political issues, does not deliver useful results, and, more typically, is run once (or a couple of times) but never run again. Just like Leo Tolstoy's happy families,[3] the success of a data science project is dependent on a number of factors. Successful data science projects need focus, good-quality data, the right people, the willingness to experiment with multiple models, integration into the business information technology (IT) architecture and processes, buy-in from senior management, and an organization's recognition that because the world changes, models go out of date and need to be rebuilt semiregularly. Failure in any of these areas is likely to result in a failed project. This section details the common factors that determine the success of data science projects as well as the typical reasons why data science projects fail.

Focus

Every successful data science project begins by clearly defining the problem that the project will help solve. In many ways, this step is just common sense: it is difficult for a project to be successful unless it has a clear goal. Having a well-defined goal informs the decisions regarding which data to use, what ML algorithms to use, how to evaluate the results, how the analysis and models will be used and deployed, and when the optimal time might be to go through the process again to update the analysis and models.

Data

A well-defined question can be used to define what data are needed for the project. Having a clear understanding of what data are needed helps to direct the project to where these required data are located. It also helps with defining what data are currently unavailable and hence identifies some additional projects that can look at capturing and

Every successful data
science project begins
by clearly defining
the problem that the
project will help solve.

making available these data. It is important, however, to ensure that the data used are good-quality data. Organizations may have applications that are poorly designed, a very poor data model, and staff who are not trained correctly to ensure that good data get entered. In fact, myriad factors can lead to bad-quality data in systems. Indeed, the need for good-quality data is so important that some organizations have hired people to constantly inspect the data, assess the quality of the data, and then feed back ideas on how to improve the quality of the data captured by the applications and by the people inputting the data. Without good-quality data, it is very difficult for a data science project to succeed.

When the required data are sourced, it is always important to check what data are being captured and used across an organization. Unfortunately, the approach to sourcing data taken by some data science projects is to look at what data are available in the transactional databases (and other data sources) and then to integrate and clean these data before going on to data exploration and analysis. This approach completely ignores the BI team and any data warehouse that might exist. In many organizations, the BI and data-warehouse team is already gathering, cleaning, transforming, and integrating the organization's data into one central repository. If a data warehouse already exists, then it probably contains all or most of the data required by a project. Therefore, a data warehouse can save a significant amount of time on integrating and cleaning the data. It will also have much more data than the current transactional databases contain. If the data warehouse is used, it is possible to go back a number of years, build predictive models using the historic data, roll these models through various time periods, and then measure each model's level of predictive accuracy. This process allows for the monitoring of changes in the data and how they affect the models. In addition, it is possible to monitor variations in the models that are produced by ML algorithms and how the models evolve over time. Following this kind of approach facilitates the demonstration of how the models work and behave over a number of years and helps with building up the customer's confidence in what is being done and what can be achieved. For example, in one project where five years of historical data were available in the data warehouse, it was possible to demonstrate that the company could have saved US$40 million or more over that time period. If the data warehouse had not been available or used, then it would not have been possible to demonstrate this conclusion. Finally, when a project is using personal data it is essential to ensure that the use of this data is in line with the relevant antidiscrimination and privacy regulations.

People

A successful data science project often involves a team of people with a blend of data science competencies and skills. In most organizations, a variety of people in existing

A successful data science project often involves a team of people with a blend of data science competencies and skills.

roles can and should contribute to data science projects: people working with databases, people who work with the ETL process, people who perform data integration, project managers, business analysts, domain experts, and so on. But organizations often still need to hire data science specialists—that is, people with the skills to work with big data, to apply ML, and to frame real-world problems in terms of data-driven solutions. Successful data scientists are willing and able to work and communicate with the management team, end users, and all involved to show and explain what and how data science can support their work. It is difficult to find people who have both the required technical skill set and the ability to communicate and work with people across an organization. However, this blend is crucial to the success of data science projects in most organizations.

Models

It is import to experiment with a variety of ML algorithms to discover which works best with the data sets. All too often in the literature, examples are given of cases where only one ML algorithm was used. Maybe the authors are discussing the algorithm that worked best for them or that is their favorite. Currently there is a great deal of interest in the use of neural networks and deep learning. Many other algorithms can be used, however, and these alternatives should be considered and tested. Furthermore, for data science projects based in the EU, the General Data Protection Regulations, which go into effect in April 2018, may become a factor in determining the selection of algorithms and model. A potential side effect of these regulations is that an individual's "right to explanation" in relation to automated decision processes that affect them may limit the use in some domains of complex models that are difficult to interpret and explain (such as deep neural network models).

Integration with the Business

When the goal of a data science project is being defined, it is vital also to define how the outputs and results of the project will be deployed within the organization's IT architecture and business processes. Doing so involves identifying where and how the model is to be integrated within existing systems and how the generated results will be used by the system end users or if the results will be fed into another process. The more automated this process is, the quicker the organization can respond to its customers' changing profile, thereby reducing costs and increasing potential profits. For example, if a customer-risk model is built for the loan process in a bank, it should be built into the front-end system that captures the loan application by the customer. That way, when the bank employee is entering the loan application, she can be given live feedback by the model. The employee can then use this live feedback to address any issues with the customer. Another example is fraud detection. It can take four to six weeks to identify a potential fraud case that needs investigation. By using data science and building it into transaction-monitoring systems, organizations can now detect potential fraud cases in near real time. By automating and integrating data-driven models, quicker response times are achieved, and actions can be taken at the right time. If the outputs and models created by a project are not integrated into the business processes, then these outputs will not be used, and, ultimately, the project will fail.

Buy-in

For most projects in most organizations, support by senior management is crucial to the success of many data science projects. However, most senior IT managers are very focused on the here and now: keeping the lights on, making sure their day-to-day applications are up and running, making sure the backups and recovery processes are in place (and tested), and so on. Successful data science projects are sponsored by senior business managers (rather than by an IT manager) because the former are focused not on the technology but on the processes involved in the data science project and how the outputs of the data science project can be used to the organization's advantage. The more focused a project sponsor is on these factors, the more successful the project will be.

For an organization to reap long-term benefits, it needs to build its capacity to execute data science projects often and to use the outputs of these projects.

He or she will then act as the key to informing the rest of the organization about the project and selling it to them. But even when data science has a senior manager as an internal champion, a data science strategy can still fail in the long term if the initial data science project is treated as a box-ticking exercise. The organization should not view data science as a one-off project. For an organization to reap long-term benefits, it needs to build its capacity to execute data science projects often and to use the outputs of these projects. It takes long-term commitment from senior management to view data science as a strategy.

Iteration

Most data science projects will need to be updated and refreshed on a semiregular basis. For each new update or iteration, new data can be added, new updates can be added, maybe new algorithms can be used, and so on. The frequency of these iterations will vary from project to project; it could be daily or quarterly or biannually or annually. Checks should be built into the productionalized data science outputs to detect when models need updating (see Kelleher, Mac Namee, and D'Arcy 2015 for an explanation of how to use a stability index to identify when a model should be updated).

Final Thoughts

Humans have always abstracted from the world and tried to understand it by identifying patterns in their experiences of it. Data science is the latest incarnation of this pattern-seeking behavior. However, although data science has a long history, the breadth of its impact on modern life is without precedent. In modern societies, the words *precision, smart, targeted*, and *personalized* are often indicative of data science projects: *precision medicine, precision policing, precision agriculture, smart cities, smart transport, targeted advertising, personalized entertainment*. The common factor across all these areas of human life is that decisions have to be made: What treatment should we use for this patient? Where should we allocate our policing resources? How much fertilizer should we spread? How many high schools do we need to build in the next four years? Who should we send this advertisement to? What movie or book should we recommend to this person? The power of data science to help with decision making is driving its adoption. Done well, data science can provide *actionable insight* that leads to better decisions and ultimately better outcomes.

Data science, in its modern guise, is driven by big data, computer power, and human ingenuity from a number of fields of scientific endeavor (from data mining and database research to machine learning). This book has tried to provide an overview of the fundamental ideas and concepts required to understand data science. The CRISP-DM project life cycle makes the data science process explicit and provides a structure for the data science journey from data to wisdom: understand the problem, prepare the data, use ML to extract patterns and create models, use the models to get actionable insight. The book also touches on some of the ethical concerns relating to individual privacy in a data science world. People have genuine and well-founded concerns that data science has the potential to be used by governments and vested interests to manipulate our behaviors and police our actions. We, as individuals, need to develop informed opinions about what type of a data world we want to live in and to think about the laws we want our societies to develop in order to steer the use of data science in appropriate directions. Despite the ethical concerns we may have around data science, the genie is already very much out of the bottle: data science is having and will continue to have significant effects on our daily lives. When used appropriately, it has the potential to improve our lives. But if we want the organizations we work with, the communities we live in, and the families we share our lives with to benefit from data science, we need to understand and explore what data science is, how it works, and what it can (and can't) do. We hope this book has given you the essential foundations you need to go on this journey.

Analytics Base Table

A table in which each row contains the data relating to a specific instance and each column describes the values of a particular attribute for each instance. These data are the basic input to data-mining and machine-learning algorithms.

Anomaly Detection

Searching for and identifying examples of atypical data in a data set. These nonconforming cases are often referred to as *anomalies* or *outliers*. This process is often used in analyzing financial transactions to identify potential fraudulent activities and to trigger investigations.

Association-Rule Mining

An unsupervised data-analysis technique that looks to find groups of items that frequently co-occur together. The classic use case is market-basket analysis, where retail companies try to identify sets of items that are purchased together, such as the hot dogs, ketchup, and beer.

Attribute

Each instance in a data set is described by a number of attributes (also known as *features* or *variables*). An attribute captures one piece of information relating to an instance. An attribute can be either raw or derived.

Backpropagation

The backpropagation algorithm is an ML algorithm used to train neural networks. The algorithm calculates for each neuron in a network the contribution the neuron makes to the error of the network. Using this error calculation for each neuron it is possible to update the weights on the inputs to each neuron so as to reduce the overall error of the network. The backpropagation algorithm is so named because it works in a two stage process. In the first stage an instance is input to the network and the information flows forward through the network until the network generates a prediction for that instance. In the second stage the error of the network on that instance is calculated by comparing the network's prediction to the correct output for that instance (as specified by the training data) and then this error is then shared back (or backpropagated) through the neurons in the network on a layer by layer basis beginning at the output layer.

Big Data
Big data are often defined in terms of the three Vs: the extreme volume of data, the variety of the data types, and the velocity at which the data must be processed.

Captured Data
Data that are captured through a direct measurement process that is designed to gather the data. Contrast with **exhaust data**.

Classification
The task of predicting a value for a target attribute of an instance based on the values of a set of input attributes, where the target attribute is a nominal or ordinal data type.

Clustering
Identifying groups of similar instances in a data set.

Correlation
The strength of association between two attributes.

Cross Industry Standard Process for Data Mining (CRISP-DM)
The CRISP-DM defines a standard life cycle for a data-mining project. The life cycle is often adopted for data science projects.

Data   In its most basic form, a piece of data is an abstraction (or measurement) from a real-world entity (person, object, or event).

Data Analysis
Any process for extracting useful information from data. Types of data analysis include data visualization, summary statistics, correlation analysis, and modeling using machine learning.

Database
A central repository of data. The most common database structure is a relational database, which stores data in tables with a structure of one row per

instance and one column per attribute. This representation is ideal for storing data with a clear structure that can be decomposed into natural attributes.

Data Mining

The process of extracting useful patterns from a data set to solve a well-defined problem. CRISP-DM defines the standard life cycle for a data-mining project. Closely related to data science but in general not as broad in scope.

Data Science

An emerging field that integrates a set of problem definitions, algorithms, and processes that can be used to analyze data so as to extract actionable insight from (large) data sets. Closely related to the field of data mining but broader in scope and concern. Deals with both structured and unstructured (big) data and encompasses principles from a range of fields, including machine learning, statistics, data ethics and regulation, and high-performance computing.

Data Set

A collection of data relating to a set of instances, with each instance described in terms of a set of attributes. In its most basic form, a data set is organized in an $n * m$ matrix, where n is the number of instances (rows) and m is the number of attributes (columns).

Data Warehouse

A centralized repository containing data from a range of sources across an organization. The data are structured to support summary reports from the aggregated data. *Online analytical processing* (OLAP) is the term used to describe the typical operations on a data warehouse.

Decision Tree

A type of prediction model that encodes *if-then-else* rules in a tree structure. Each node in the tree defines one attribute to test, and a path from the root node to a terminating leaf node defines a sequence of tests that an instance must pass for the label of the terminating node to be predicted for that instance.

Deep Learning

A deep-learning model is a neural network that has multiple (more than two) layers of hidden units (or neurons). Deep networks are deep in terms of the number of layers of neurons in the network. Today many deep networks have tens to hundreds of layers. The power of deep-learning models comes from the ability of the neurons in the later layers to learn useful attributes derived from attributes that were themselves learned by the neurons in the earlier layers.

Derived Attribute

An attribute whose value is generated by applying a function to other data rather than a direct measurement taken from the entity. An attribute that describes an average value in a population is an example of a derived attribute. Contrast with **raw attribute.**

DIKW Pyramid

A model of the structural relationships between *data, information, knowledge,* and *wisdom*. In the DIKW pyramid, data precedes information, which precedes knowledge, which precedes wisdom.

Exhaust Data

Data that are a by-product of a process whose primary purpose is something other than data capture. For example, for every image shared, tweeted, retweeted, or liked, a range of exhaust data is generated: who shared, who viewed, what device was used, what time of day, and so on. Contrast with **captured data**.

Extraction, Transformation, and Load (ETL)

ETL is the term used to describe the typical processes and tools used to support the mapping, merging, and movement of data between databases.

Hadoop

Hadoop is an open-source framework developed by the Apache Software Foundation that is designed for the processing of big data. It uses distributed storage and processing across clusters of commodity hardware.

High-Performance Computing (HPC)

The field of HPC focuses on designing and implementing frameworks to connect large number of computers together so that the resulting computer cluster can store and process large amounts of data efficiently.

In-Database Machine Learning

Using machine-learning algorithms that are built into the database solution. The benefit of in-database machine learning is that it reduces the time spent on moving data in and out of databases for analysis.

Instance
Each row in a data set contains the information relating to one instance (also known as an *example*, *entity*, *case*, or *record*).

Internet of Things
The internetworking of physical devices and sensors so that these devices can share information. Includes the field of machine-to-machine communication, which develops systems that enable machines not only to share information but also to react to this information and trigger actions without human involvement.

Linear Regression
When a linear relationship is assumed in a **regression analysis**, the analysis is called *linear regression*. A popular type of prediction model used to estimate the value of a numeric target attribute based on a set of numeric input attributes.

Machine Learning (ML)
The field of computer science research that focuses on developing and evaluating algorithms that can extract useful patterns from data sets. A machine-learning algorithm takes a **data set** as input and returns a **model** that encodes the patterns the algorithm extracted from the data.

Massively Parallel Processing Database (MPP)
In an MPP database, data is partitioned across multiple servers, and each server can process the data on that server locally and independently.

Metadata
Data describing the structures and properties of other data—for example, a time stamp that describes when a piece of data was collected. Metadata are one of the most common types of **exhaust data**.

Model
In the context of machine learning, a model is a representation of a pattern extracted using machine learning from a data set. Consequently, models are trained, fitted to a data set, or created by running a machine learning algorithm on a data set. Popular model representations include **decision trees** and **neural networks**. A **prediction** model defines a mapping (or function) from a set of input attributes to a value for a **target attribute**. Once a model has been created, it can then be applied to new instances from the domain. For example,

in order to train a spam filter model, we would apply a machine learning algorithm to a data set of historic emails that have been labeled as spam or not spam. Once the model has been trained it can be used to label (or filter) new emails that were not in the original data set.

Neural Network
A type of machine-learning model that is implemented as a network of simple processing units called **neurons**. It is possible to create a variety of different types of neural networks by modifying the topology of the neurons in the network. A feed-forward, fully connected neural network is a very common type of network that can be trained using **backpropagation**.

Neuron
A neuron takes a number of input values (or activations) as input and maps these values to a single output activation. This mapping is typically implemented by applying a multi-input linear-regression function to the inputs and then pushing the result of this regression function through a nonlinear activation function, such as the logistic or tanh function.

Online Analytical Processing (OLAP)
OLAP operations generate summaries of historic data and aggregate data from multiple sources. OLAP operations are designed to generate report-type summaries and enable users to slice, dice, and pivot data in a data warehouse using a predefined set of dimensions on the data, such as sales by stores, sale by quarter, and so on. Contrast with **Online Transaction Processing (OLTP)**.

Online Transaction Processing (OLTP)
OLTPs are designed for short online data transactions (such as INSERT, DE-LETE, UPDATE, etc.) with an emphasis on fast query processing and maintaining data integrity in a multi-access environment. Contrast with **OLAP** systems, which are designed for more complex operations on historic data.

Operational Data Store (ODS)
An ODS system integrates operational or transactional data from multiple systems to support operational reporting.

Prediction
In the context of data science and machine learning, the task of estimating the value of a target attribute for a given instance based on the values of other attributes (or input attributes) for that instance.

Raw Attribute

An abstraction from an entity that is a direct measurement taken from the entity—for example, a person's height. Contrast with **derived attribute**.

Regression Analysis

Estimates the expected (or average) value of a numeric target attribute when all the input attribute values are fixed. Regression analysis assumes a parameterized mathematical model of the hypothesized relationship between the inputs and output known as a *regression function*. A regression function may have multiple parameters, and the focus of regression analysis is to find the correct settings for these parameters.

Relational Database Management System (RDBMS)

Database management systems based on Edgar F. Codd's relational data model. Relational databases store data in collection of tables where each table has a structure of one row per instance and one column per attribute. Links between tables can be created by having key attributes appear in multiple tables. This structure is suited for SQL queries which define operations on the data in the tables.

Smart City

Smart-city projects generally try to integrate real-time data from many different data sources into a single data hub, where they are analyzed and used to inform city-management and planning decisions.

Structured Data

Data that can be stored in a table. Every instance in the table has the same set of attributes. Contrast with **unstructured data**.

Structured Query Language (SQL)

An international standard for defining database queries.

Supervised Learning

A form of machine learning in which the goal is to learn a function that maps from a set of input attribute values for an instance to an estimate of the missing value for the target attribute of the same instance.

Target Attribute

In a prediction task, the attribute that the prediction model is trained to estimate the value of.

Transactional Data
Event information, such as the sale of an item, the issuing of an invoice, the delivery of goods, credit card payment, and so on.

Unstructured Data
A type of data where each instance in the data set may have its own internal structure; that is, the structure is not necessarily the same in every instance. For example, text data are often unstructured and require a sequence of operations to be applied to them in order to extract a structured representation for each instance.

Unsupervised Learning
A form of machine learning in which the goal is to identify regularities in the data. These regularities may include clusters of similar instances within the data or regularities between attributes. In contrast to **supervised learning**, in unsupervised learning no target attribute is defined in the data set.

Chapter 1

1. Quote taken from the call for participation sent out for the KDD workshop in 1989.

2. Some practitioners do distinguish between data mining and KDD by viewing data mining as a subfield of KDD or a particular approach to KDD.

3. For a recent review of this debate, see *Battle of the Data Science Venn Diagrams* (Taylor 2016).

4. For more on the Cancer Moonshot Initiative, see https://www.cancer.gov/research/key-initiatives.

5. For more on the All of Us program in the Precision Medicine Initiative, see https://allofus.nih.gov.

6. For more on the Police Data Initiative, see https://www.policedatainitiative.org.

7. For more on AlphaGo, see https://deepmind.com/research/alphago.

Chapter 2

1. Although many data sets can be described as a flat $n * m$ matrix, in some scenarios the data set is more complex: for example, if a data set describes the evolution of multiple attributes through time, then each time point in the data set will be represented by a two-dimensional flat $n * m$ matrix, listing the state of the attributes at that point in time, but the overall data set will be three dimensional, where time is used to link the two-dimensional snapshots. In these contexts, the term *tensor* is sometimes used to generalize the *matrix* concept to higher dimensions.

2. This example is inspired by an example in Han, Kamber, and Pei 2011.

Chapter 3

1. See Storm website, at http://storm.apache.org.

Chapter 4

1. This subheading, Correlations Are Not Causations, but Some Are Useful, is inspired by George E. P. Box's (1979) observation, "Essentially, all models are wrong, but some are useful."

2. For a numeric target, the average is the most common measure of central tendency, and for nominal or ordinal data the mode (or most frequently occurring value is the most common measure of central tendency).

3. We are using a more complex notation here involving ω_0 and ω_1 because a few paragraphs later we expand this function to include more than one input attribute, so the subscripted variables are useful notations when dealing with multiple inputs.

4. A note of caution: the numeric values reported here should be taken as illustrative only and not interpreted as definitive estimates of the relationship between BMI and likelihood of diabetes.

5. In general, neural networks work best when the inputs have similar ranges. If there are large differences in the ranges of input attributes, the attributes with the much larger values tend to dominate the processing of the network. To avoid this, it is best to normalize the input attributes so that they all have similar ranges.

6. For the sake of simplicity, we have not included the weights on the connections in figures 14 and 15.

7. Technically, the backpropagation algorithm uses the chain rule from calculus to calculate the derivative of the error of the network with respect to each weight for each neuron in the network, but for this discussion we will pass over this distinction between the error and the derivative of the error for the sake of clarity in explaining the essential idea behind the backpropagation algorithm.

8. No agreed minimum number of hidden layers is required for a network to be considered "deep," but some people would argue that even two layers are enough to be deep. Many deep networks have tens of layers, but some networks can have hundreds or even thousands of layers.

9. For an accessible introduction to RNNs and their natural-language processing, see Kelleher 2016.

10. Technically, the decrease in error estimates is known as the *vanishing-gradient problem* because the gradient over the error surface disappears as the algorithm works back through the network.

11. The algorithm also terminates on two corner cases: a branch ends up with no instances after the data set is split up, or all the input attributes have already been used at nodes between the root node and the branch. In both cases, a terminating node is added and is labeled with the majority value of the target attribute at the parent node of the branch.

12. For an introduction to entropy and its use in decision-tree algorithms, see Kelleher, Mac Namee, and D'Arcy 2015 on information-based learning.

13. See Burt 2017 for an introduction to the debate on the "right to explanation."

Chapter 5

1. A customer-churn case study in Kelleher, Mac Namee, and D'Arcy 2015 provides a longer discussion of the design of attributes in propensity models.

Chapter 6

1. Behavioral targeting uses data from users' online activities—sites visited, clicks made, time spent on a site, and so on—and predictive modeling to select the ads shown to the user.

2. The EU Privacy and Electronic Communications Directive (2002/58/EC).

3. For example, some expectant women explicitly tell retailers that they are pregnant by registering for promotional new-mother programs at the stores.

4. For more on PredPol, see http://www.predpol.com.

5. A Panopticon is an eighteenth-century design by Jeremy Bentham for institutional buildings, such as prisons and psychiatric hospitals. The defining characteristic of a Panopticon was that the staff could observe the inmates without the inmates' knowledge. The underlying idea of this design was that the inmates were forced to act as though they were being watched at all times.

6. As distinct from digital footprint.

7. Civil Rights Act of 1964, Pub. L. 88-352, 78 Stat. 241, at https://www.gpo.gov/fdsys/pkg/STATUTE-78/pdf/STATUTE-78-Pg241.pdf.

8. Americans with Disabilities Act of 1990, Pub. L. 101-336, 104 Stat. 327, at https://www.gpo.gov/fdsys/pkg/STATUTE-104/pdf/STATUTE-104-Pg327.pdf.

9. The Fair Information Practice Principles are available at https://www.dhs.gov/publication/fair-information-practice-principles-fipps.

10. Senate of California, SB-568 Privacy: Internet: Minors, Business and Professions Code, Relating to the Internet, vol. division 8, chap. 22.1 (commencing with sec. 22580) (2013), at https://leginfo.legislature.ca.gov/faces/billNavClient.xhtml?bill_id=201320140SB568.

Chapter 7

1. For more on the SmartSantander project in Spain, see http://smartsantander.eu.

2. For more on the TEPC's projects, see http://www.tepco.co.jp/en/press/corp-com/release/2015/1254972_6844.html.

3. Leo Tolstoy's book *Anna Karenina* (1877) begins: "All happy families are alike; each unhappy family is unhappy in its own way." Tolstoy's idea is that to be happy, a family must be successful in a range of areas (love, finance, health, in-laws), but failure in any of these areas will result in unhappiness. So all happy families are the same because they are successful in all areas, but unhappy families can be unhappy for many different combinations of reasons.

FURTHER READINGS

About Data and Big Data

Davenport, Thomas H. *Big Data at Work: Dispelling the Myths, Uncovering the Opportunities*. Cambridge, MA: Harvard Business Review, 2014.

Harkness, Timandra. *Big Data: Does Size Matter?* New York: Bloomsbury Sigma, 2016.

Kitchin, Rob. *The Data Revolution: Big Data, Open Data, Data Infrastructures, and Their Consequences*. Los Angeles: Sage, 2014.

Mayer-Schönberger, Viktor, and Kenneth Cukier. *Big Data: A Revolution That Will Transform How We Live, Work, and Think*. Boston: Eamon Dolan/Mariner Books, 2014.

Pomerantz, Jeffrey. *Metadata*. Cambridge, MA: MIT Press, 2015.

Rudder, Christian. *Dataclysm: Who We Are (When We Think No One's Looking)*. New York: Broadway Books, 2014.

About Data Science, Data Mining, and Machine Learning

Kelleher, John D., Brian Mac Namee, and Aoife D'Arcy. *Fundamentals of Machine Learning for Predictive Data Analytics*. Cambridge, MA: MIT Press, 2015.

Linoff, Gordon S., and Michael J. A. Berry. *Data Mining Techniques: For Marketing, Sales, and Customer Relationship Management*. Indianapolis, IN: Wiley, 2011.

Provost, Foster, and Tom Fawcett. *Data Science for Business: What You Need to Know about Data Mining and Data-Analytic Thinking*. Sebastopol, CA: O'Reilly Media, 2013.

About Privacy, Ethics, and Advertising

Dwork, Cynthia, and Aaron Roth. 2014. "The Algorithmic Foundations of Differential Privacy." *Foundations and Trends® in Theoretical Computer Science* 9 (3–4): 211–407.

Nissenbaum, Helen. *Privacy in Context: Technology, Policy, and the Integrity of Social Life*. Stanford, CA: Stanford Law Books, 2009.

Solove, Daniel J. *Nothing to Hide: The False Tradeoff between Privacy and Security*. New Haven, CT: Yale University Press, 2013.

Turow, Joseph. *The Daily You: How the New Advertising Industry Is Defining Your Identity and Your Worth*. New Haven, CT: Yale University Press, 2013.

REFERENCES

Anderson, Chris. 2008. *The Long Tail: Why the Future of Business Is Selling Less of More*. Rev. ed. New York: Hachette Books.

Baldridge, Jason. 2015. "Machine Learning and Human Bias: An Uneasy Pair." *TechCrunch*, August 2. http://social.techcrunch.com/2015/08/02/machine-learning-and-human-bias-an-uneasy-pair.

Barry-Jester, Anna Maria, Ben Casselman, and Dana Goldstein. 2015. "Should Prison Sentences Be Based on Crimes That Haven't Been Committed Yet?" *FiveThirtyEight*, August 4. https://fivethirtyeight.com/features/prison-reform-risk-assessment.

Batty, Mike, Arun Tripathi, Alice Kroll, Peter Wu Cheng-sheng, David Moore, Chris Stehno, Lucas Lau, Jim Guszcza, and Mitch Katcher. 2010. "Predictive Modeling for Life Insurance: Ways Life Insurers Can Participate in the Business Analytics Revolution." Society of Actuaries. https://www.soa.org/files/pdf/research-pred-mod-life-batty.pdf.

Beales, Howard. 2010. "The Value of Behavioral Targeting." Network Advertising Initiative. http://www.networkadvertising.org/pdfs/Beales_NAI_Study.pdf.

Berk, Richard A., and Justin Bleich. 2013. "Statistical Procedures for Forecasting Criminal Behavior." *Criminology & Public Policy* 12 (3): 513–544.

Box, George E. P. 1979. "Robustness in the Strategy of Scientific Model Building," in *Robustness in Statistics*, ed. R. L. Launer and G. N. Wilkinson, 201–236. New York: Academic Press.

Breiman, Leo. 2001. "Statistical Modeling: The Two Cultures (with Comments and a Rejoinder by the Author)." *Statistical Science* 16 (3): 199–231. doi:10.1214/ss/1009213726.

Brown, Meta S. 2014. *Data Mining for Dummies*. New York: Wiley. http://www.wiley.com/WileyCDA/WileyTitle/productCd-1118893174,subjectCd-STB0.html.

Brynjolfsson, Erik, Lorin M. Hitt, and Heekyung Hellen Kim. 2011. "Strength in Numbers: How Does Data-Driven Decisionmaking Affect Firm Performance?"

SSRN Scholarly Paper ID 1819486. Social Science Research Network, Rochester, NY. https://papers.ssrn.com/abstract=1819486.

Burt, Andrew. 2017. "Is There a 'Right to Explanation' for Machine Learning in the GDPR?" https://iapp.org/news/a/is-there-a-right-to-explanation-for-machine-learning-in-the-gdpr.

Buytendijk, Frank, and Jay Heiser. 2013. "Confronting the Privacy and Ethical Risks of Big Data." *Financial Times*, September 24. https://www.ft.com/content/105e30a4-2549-11e3-b349-00144feab7de.

Carroll, Rory. 2013. "Welcome to Utah, the NSA's Desert Home for Eavesdropping on America." *Guardian*, June 14. https://www.theguardian.com/world/2013/jun/14/nsa-utah-data-facility.

Cavoukian, Ann. 2013. "Privacy by Design: The 7 Foundation Principles (Primer)." Information and Privacy Commissioner, Ontario, Canada. https://www.ipc.on.ca/wp-content/uploads/2013/09/pbd-primer.pdf.

Chapman, Pete, Julian Clinton, Randy Kerber, Thomas Khabaza, Thomas Reinartz, Colin Shearer, and Rudiger Wirth. 1999. "CRISP-DM 1.0: Step-by-Step Data Mining Guide." ftp://ftp.software.ibm.com/software/analytics/spss/support/Modeler/Documentation/14/UserManual/CRISP-DM.pdf.

Charter of Fundamental Rights of the European Union. 2000. *Official Journal of the European Communities* C (364): 1–22.

Cleveland, William S. 2001. "Data Science: An Action Plan for Expanding the Technical Areas of the Field of Statistics." *International Statistical Review* 69 (1): 21–26. doi:10.1111/j.1751-5823.2001.tb00477.x.

Clifford, Stephanie. 2012. "Supermarkets Try Customizing Prices for Shoppers." *New York Times*, August 9. http://www.nytimes.com/2012/08/10/business/supermarkets-try-customizing-prices-for-shoppers.html.

Council of the European Union and European Parliament. 1995. "95/46/EC of the European Parliament and of the Council of 24 October 1995 on the Protection of Individuals with Regard to the Processing of Personal Data and on the Free Movement of Such Data." *Official Journal of the European Community* L 281:38-1995): 31–50.

Council of the European Union and European Parliament. 2016. "General Data Protection Regulation of the European Council and Parliament." *Official*

Journal of the European Union L 119: 1–2016. http://ec.europa.eu/justice/data-protection/reform/files/regulation_oj_en.pdf.

CrowdFlower. 2016. *2016 Data Science Report*. http://visit.crowdflower.com/rs/416-ZBE-142/images/CrowdFlower_DataScienceReport_2016.pdf.

Datta, Amit, Michael Carl Tschantz, and Anupam Datta. 2015. "Automated Experiments on Ad Privacy Settings." *Proceedings on Privacy Enhancing Technologies* 2015 (1): 92–112.

DeZyre. 2015. "How Big Data Analysis Helped Increase Walmart's Sales Turnover." May 23. https://www.dezyre.com/article/how-big-data-analysis-helped-increase-walmarts-sales-turnover/109.

Dodge, Martin, and Rob Kitchin. 2007. "The Automatic Management of Drivers and Driving Spaces." *Geoforum* 38 (2): 264–275.

Dokoupil, Tony. 2013. "'Small World of Murder': As Homicides Drop, Chicago Police Focus on Social Networks of Gangs." *NBC News*, December 17. http://www.nbcnews.com/news/other/small-world-murder-homicides-drop-chicago-police-focus-social-networks-f2D11758025.

Duhigg, Charles. 2012. "How Companies Learn Your Secrets." *New York Times*, February 16. http://www.nytimes.com/2012/02/19/magazine/shopping-habits.html.

Dwork, Cynthia, and Aaron Roth. 2014. "The Algorithmic Foundations of Differential Privacy." *Foundations and Trends® in Theoretical Computer Science* 9 (3–4): 211–407.

Eliot, T. S. 1934 [1952]. "Choruses from 'The Rock.'" In *T. S. Eliot: The Complete Poems and Plays—1909–1950*. San Diego: Harcourt, Brace and Co.

Elliott, Christopher. 2004. "BUSINESS TRAVEL; Some Rental Cars Are Keeping Tabs on the Drivers." *New York Times*, January 13. http://www.nytimes.com/2004/01/13/business/business-travel-some-rental-cars-are-keeping-tabs-on-the-drivers.html.

Eurobarometer. 2015. "Data Protection." Special Eurobarometer 431. http://ec.europa.eu/COMMFrontOffice/publicopinion/index.cfm/Survey/index#p=1&instruments=SPECIAL.

European Commission. 2012. "Commission Proposes a Comprehensive Reform of the Data Protection Rules—European Commission." January 25. http://ec.europa.eu/justice/newsroom/data-protection/news/120125_en.htm.

European Commission. 2016. "The EU-U.S. Privacy Shield." December 7. http://ec.europa.eu/justice/data-protection/international-transfers/eu-us -privacy-shield/index_en.htm.

Federal Trade Commission. 2012. Protecting Consumer Privacy in an Era of Rapid Change. Washington, DC: Federal Trade Commission. https://www .ftc.gov/sites/default/files/documents/reports/federal-trade-commission -report-protecting-consumer-privacy-era-rapid-change-recommendations/12 0326privacyreport.pdf.

Few, Stephen. 2012. *Show Me the Numbers: Designing Tables and Graphs to Enlighten*. 2nd ed. Burlingame, CA: Analytics Press.

Goldfarb, Avi, and Catherine E. Tucker. 2011. Online Advertising, Behavioral Targeting, and Privacy. *Communications of the ACM* 54 (5): 25–27.

Gorner, Jeremy. 2013. "Chicago Police Use Heat List as Strategy to Prevent Violence." *Chicago Tribune*, August 21. http://articles.chicagotribune.com/ 2013-08-21/news/ct-met-heat-list-20130821_1_chicago-police-commander -andrew-papachristos-heat-list.

Hall, Mark, Ian Witten, and Eibe Frank. 2011. *Data Mining: Practical Machine Learning Tools and Techniques*. Amsterdam: Morgan Kaufmann.

Han, Jiawei, Micheline Kamber, and Jian Pei. 2011. *Data Mining: Concepts and Techniques*. 3rd ed. Haryana, India: Morgan Kaufmann.

Harkness, Timandra. 2016. *Big Data: Does Size Matter?* New York: Bloomsbury Sigma.

Henke, Nicolaus, Jacques Bughin, Michael Chui, James Manyika, Tamim Saleh, and Bill Wiseman. 2016. *The Age of Analytics: Competing in a Data-Driven World*. Chicago: McKinsey Global Institute. http://www.mckinsey.com/ business-functions/mckinsey-analytics/our-insights/the-age-of-analytics -competing-in-a-data-driven-world.

Hill, Shawndra, Foster Provost, and Chris Volinsky. 2006. Network-Based Marketing: Identifying Likely Adopters via Consumer Networks. *Statistical Science* 21 (2): 256–276. doi:10.1214/088342306000000222.

Hunt, Priscillia, Jessica Saunders, and John S. Hollywood. 2014. *Evaluation of the Shreveport Predictive Policing Experiment*. Santa Monica, CA: Rand Corporation. http://www.rand.org/pubs/research_reports/RR531.

Innes, Martin. 2001. Control Creep. *Sociological Research Online* 6 (3). https://ideas.repec.org/a/sro/srosro/2001-45-2.html.

Kelleher, John D. 2016. "Fundamentals of Machine Learning for Neural Machine Translation." In *Proceedings of the European Translation Forum*, 1–15. Brussels: European Commission Directorate-General for Translation. https://tinyurl.com/RecurrentNeuralNetworks.

Kelleher, John D., Brian Mac Namee, and Aoife D'Arcy. 2015. *Fundamentals of Machine Learning for Predictive Data Analytics*. Cambridge, MA: MIT Press.

Kerr, Aphra. 2017. *Global Games: Production, Circulation, and Policy in the Networked Era*. New York: Routledge.

Kitchin, Rob. 2014a. *The Data Revolution: Big Data, Open Data, Data Infrastructures, and Their Consequences*. Los Angeles: Sage.

Kitchin, Rob. 2014b. "The Real-Time City? Big Data and Smart Urbanism." *GeoJournal* 79 (1): 1–14. doi:10.1007/s10708-013-9516-8.

Koops, Bert-Jaap. 2011. "Forgetting Footprints, Shunning Shadows: A Critical Analysis of the 'Right to Be Forgotten' in Big Data Practice." Tilburg Law School Legal Studies Research Paper no. 08/2012. *SCRIPTed* 8 (3): 229–56. doi:10.2139/ssrn.1986719.

Korzybski, Alfred. 1996. "On Structure." In *Science and Sanity: An Introduction to Non-Aristotelian Systems and General Semantics*, CD-ROM, ed. Charlotte Schuchardt-Read. Englewood, NJ: Institute of General Semantics. http://esgs.free.fr/uk/art/sands.htm.

Kosinski, Michal, David Stillwell, and Thore Graepel. 2013. "Private Traits and Attributes Are Predictable from Digital Records of Human Behavior." *Proceedings of the National Academy of Sciences of the United States of America* 110 (15): 5802–5805. doi:10.1073/pnas.1218772110.

Le Cun, Yann. 1989. *Generalization and Network Design Strategies*. Technical Report CRG-TR-89-4. Toronto: University of Toronto Connectionist Research Group.

Levitt, Steven D., and Stephen J. Dubner. 2009. *Freakonomics: A Rogue Economist Explores the Hidden Side of Everything*. New York: William Morrow Paperbacks.

Lewis, Michael. 2004. *Moneyball: The Art of Winning an Unfair Game*. New York: Norton.

Linoff, Gordon S., and Michael J.A. Berry. 2011. *Data Mining Techniques: For Marketing, Sales, and Customer Relationship Management.* Indianapolis, IN: Wiley.

Manyika, James, Michael Chui, Brad Brown, Jacques Bughin, Richard Dobbs, Charles Roxburgh, and Angela Hung Byers. 2011. *Big Data: The Next Frontier for Innovation, Competition, and Productivity.* Chicago: McKinsey Global Institute. http://www.mckinsey.com/business-functions/digital-mckinsey/our-insights/big-data-the-next-frontier-for-innovation.

Marr, Bernard. 2015. *Big Data: Using SMART Big Data, Analytics, and Metrics to Make Better Decisions and Improve Performance.* Chichester, UK: Wiley.

Mayer, J. R., and J. C. Mitchell. 2012. "Third-Party Web Tracking: Policy and Technology." In *2012 IEEE Symposium on Security and Privacy*, 413–27. Piscataway, NJ: IEEE. doi:10.1109/SP.2012.47.

Mayer, Jonathan, and Patrick Mutchler. 2014. "MetaPhone: The Sensitivity of Telephone Metadata." *Web Policy*, March 12. http://webpolicy.org/2014/03/12/metaphone-the-sensitivity-of-telephone-metadata.

Mayer-Schönberger, Viktor, and Kenneth Cukier. 2014. *Big Data: A Revolution That Will Transform How We Live, Work, and Think.* Reprint. Boston: Eamon Dolan/Mariner Books.

McMahan, Brendan, and Daniel Ramage. 2017. "Federated Learning: Collaborative Machine Learning without Centralized Training Data." *Google Research Blog*, April. https://research.googleblog.com/2017/04/federated-learning-collaborative.html.

Nilsson, Nils. 1965. *Learning Machines: Foundations of Trainable Pattern-Classifying Systems.* New York: McGraw-Hill.

Oakland Privacy Working Group. 2015. "PredPol: An Open Letter to the Oakland City Council." June 25. https://www.indybay.org/newsitems/2015/06/25/18773987.php.

Organisation for Economic Co-operation and Development (OECD). 1980. *Guidelines on the Protection of Privacy and Transborder Flows of Personal Data.* Paris: OECD. https://www.oecd.org/sti/ieconomy/oecdguidelinesontheprotectionofprivacyandtransborderflowsofpersonaldata.htm.

Organisation for Economic Co-operation and Development (OECD). 2013. *2013 OECD Privacy Guidelines*. Paris: OECD. https://www.oecd.org/internet/ ieconomy/privacy-guidelines.htm.

O'Rourke, Cristín, and Aphra Kerr. 2017. "Privacy Schield for Whom? Key Actors and Privacy Discourse on Twitter and in Newspapers." In "Redesigning or Redefining Privacy?," special issue of *Westminster Papers in Communication and Culture* 12 (3): 21–36. doi:http://doi.org/ 10.16997/wpcc.264.

Pomerantz, Jeffrey. 2015. *Metadata*. Cambridge, MA: MIT Press. https:// mitpress.mit.edu/books/metadata-0.

Purcell, Kristen, Joanna Brenner, and Lee Rainie. 2012. "Search Engine Use 2012." Pew Research Center, March 9. http://www.pewinternet.org/2012/ 03/09/main-findings-11/.

Quinlan, J. R. 1986. "Induction of Decision Trees." *Machine Learning* 1 (1): 81–106. doi:10.1023/A:1022643204877.

Rainie, Lee, and Mary Madden. 2015. "Americans' Privacy Strategies Post-Snowden." Pew Research Center, March. http://www.pewinternet.org/files/ 2015/03/PI_AmericansPrivacyStrategies_0316151.pdf.

Rhee, Nissa. 2016. "Study Casts Doubt on Chicago Police's Secretive 'Heat List.'" *Chicago Magazine*, August 17. http://www.chicagomag.com/city-life/ August-2016/Chicago-Police-Data/.

Saunders, Jessica, Priscillia Hunt, and John S. Hollywood. 2016. "Predictions Put into Practice: A Quasi-Experimental Evaluation of Chicago's Predictive Policing Pilot." *Journal of Experimental Criminology* 12 (3): 347–371. doi:10.1007/s11292-016-9272-0.

Shmueli, Galit. 2010. "To Explain or to Predict?" *Statistical Science* 25 (3): 289–310. doi:10.1214/10-STS330.

Shubber, Kadhim. 2013. "A Simple Guide to GCHQ's Internet Surveillance Programme Tempora." *WIRED UK*, July 24. http://www.wired.co.uk/article/ gchq-tempora-101.

Silver, David, Aja Huang, Chris J. Maddison, Arthur Guez, Laurent Sifre, George van den Driessche, Julian Schrittwieser, et al. 2016. "Mastering the Game of *Go* with Deep Neural Networks and Tree Search." *Nature* 529 (7587): 484–489. doi:10.1038/nature16961.

Soldatov, Andrei, and Irina Borogan. 2012. "In Ex-Soviet States, Russian Spy Tech Still Watches You." *WIRED*, December 21. https://www.wired.com/2012/12/russias-hand.

Steinberg, Dan. 2013. "How Much Time Needs to Be Spent Preparing Data for Analysis?" http://info.salford-systems.com/blog/bid/299181/How-Much-Time-Needs-to-be-Spent-Preparing-Data-for-Analysis.

Taylor, David. 2016. "Battle of the Data Science Venn Diagrams." *KDnuggets*, October. http://www.kdnuggets.com/2016/10/battle-data-science-venn-diagrams.html.

Tufte, Edward R. 2001. *The Visual Display of Quantitative Information*. 2nd ed. Cheshire, CT: Graphics Press.

Turow, Joseph. 2013. *The Daily You: How the New Advertising Industry Is Defining Your Identity and Your Worth*. New Haven, CT: Yale University Press.

Verbeke, Wouter, David Martens, Christophe Mues, and Bart Baesens. 2011. "Building Comprehensible Customer Churn Prediction Models with Advanced Rule Induction Techniques." *Expert Systems with Applications* 38 (3): 2354–2364.

Weissman, Cale Gutherie. 2015. "The NYPD's Newest Technology May Be Recording Conversations." *Business Insider*, March 26. http://uk.businessinsider.com/the-nypds-newest-technology-may-be-recording-conversations-2015-3.

Wolpert, D. H., and W. G. Macready. 1997. "No Free Lunch Theorems for Optimization." *IEEE Transactions on Evolutionary Computation* 1 (1): 67–82. doi:10.1109/4235.585893.

JOHN D. KELLEHER is a Professor of Computer Science and the Academic Leader of the Information, Communication, and Entertainment Research Institute at the Dublin Institute of Technology. His research is supported by the ADAPT Centre, which is funded by Science Foundation Ireland (Grant 13/ RC/2106) and co-funded by the European Regional Development fund. He is the coauthor of *Fundamentals of Machine Learning for Predictive Data Analytics* (MIT Press).

BRENDAN TIERNEY is a lecturer in the School of Computing at the Dublin Institute of Technology, an Oracle ACE director, and the author of a number of books on data mining using Oracle technology.